# LARSON, BOSWELL, KANOLD, STIFF

# Passport
## to Algebra and Geometry

# Alternative Assessment

## by Cheryl A. Leech

Alternative Assessment includes information on using scoring rubrics, portfolios, error notebooks, math journals, and notebook quizzes. Also included are projects, mid-chapter partner quizzes, and individual and group assessments.

**McDougal Littell**
A HOUGHTON MIFFLIN COMPANY
Evanston, Illinois • Boston • Dallas

# CONTENTS

**ASSESSMENT**

The purpose of assessment in mathematics is to improve learning and teaching. It is imperative that assessment be used to broaden and inform, not restrict the educational process.

**NCTM STANDARDS**

With the advent of the standards for mathematics proposed by the National Council of Teachers of Mathematics (NCTM), the process of change in mathematics education has accelerated. No longer is the emphasis on computation. The emphasis is now shifting to include higher-order thinking skills, problem solving, and the ability to communicate mathematically. These new emphases do not so much require that a different type of mathematics be taught, rather that the mathematics be taught differently. These changes call for different types of assessment than the traditional paper-and-pencil tests and/or quizzes where there is only one answer and only the answer is evaluated. According to the *NCTM Curriculum and Evaluation Standards for School Mathematics,*

> *"Assessment must be more than testing; it must be a continuous, dynamic, and often informal process,"*
> (p. 203).

Alternative assessment is a means of evaluating student progress using several non-traditional assessment tools.

**PURPOSE**

According to the Mathematical Sciences Education Board (MSEB), there is a need for mathematics assessments that accomplish the following.

- Promote the development of mathematical power for all students.

- Measure the full range of mathematical knowledge, skills, and processes specified by the NCTM curriculum standards.

- Communicate to students, parents, and teachers what mathematics students already know, as well as the mathematics they have yet to learn.

**WHO USES THE INFORMATION?**

The information gathered by alternative assessment can be used in various ways. Students can use the information gathered through alternative assessment to appraise their own mathematical achievement and understanding. Teachers can use the information to make informed decisions about the instruction of their students. Administrators can use this information to evaluate the effectiveness of a mathematics program. And finally, policy makers would find this information invaluable for budget appropriations for mathematics education programs.

**ADVANTAGES**

There are several advantages to using a variety of measures of achievement.
- One type of assessment cannot serve all informational needs.
- Receiving information from multiple sources leads to more informed decision making.
- Traditional paper-and-pencil tests are incomplete measures of achievement.
- Using a variety of assessments is a more equitable measure of a student's mathematical progress. Many alternative forms of assessments have less potential for bias than traditional measures.

**GOALS**

The goals of alternative assessment are as follows.
- Find out what the student already knows.
- Evaluate the depth of the student's conceptual understanding, and his or her ability to transfer this understanding to new and different situations.
- Evaluate the student's ability to communicate mathematically his or her understanding, make mathematical connections, and to reason mathematically.
- Help plan the mathematics instruction necessary to achieve the objectives of the course.
- Report individual student progress and show growth towards mathematical maturity.
- Analyze the overall effectiveness of the mathematics instruction.

**BIBLIOGRAPHY**

Here are some  additional sources of readings about alternative assessment.

*Curriculum and Evaluation Standards for School Mathematics*, National Council of Teachers of Mathematics, 1989.

*How to Teach Math with USA Today*, USA Today, 1991.

Mumme, Judy.  *Portfolio Assessment in Mathematics*, California Mathematics Project, University of California, Santa Barbara, 1990.

Smith, Sanderson M. *Great Ideas for Teaching Math,* Portland, Maine:  J. Weston Walch Publishing, 1990.

Stenmark, Jean Kerr. "Assessment Alternatives in Mathematics," *Equals*, 1989.

**HOLISTIC SCORING**

Some teachers hesitate to use alternative assessment methods because they are not sure of how to grade the assignments. One method that many educators have found useful is called holistic scoring. Mathematics educators have long used a similar method to evaluate a student's answer by first examining the answer and then analyzing the student's work if the answer is incorrect. Holistic scoring evaluates the solution as a whole, instead of focusing on only one aspect of the solution. To reach a final score, you must first identify aspects of the work that you consider necessary to the solution. Be sure to share your list with students. Here are some aspects of the student's work that you should consider.

- Understanding the Task
- How? Quality of Approaches or Procedures
- Why? Decisions Made Along the Way
- What? Outcomes of Activities
- Language of Mathematics
- Mathematical Representations
- Clarity of Presentation

Each of these aspects is discussed below.

**UNDERSTANDING THE TASK**

*Sources of Evidence*
a. Explanation of task
b. Reasonableness of approach
c. Correctness of response, inference of understanding

*Final Rating*
1. Totally misunderstood
2. Partially understood
3. Understood
4. Generalized, applied, extended

**HOW? QUALITY OF APPROACHES OR PROCEDURES**

*Sources of Evidence*
a. Demonstrations
b. Descriptions (oral or written)
c. Drafts, scratch work, and sketches

*Final Rating*
1. Inappropriate or unworkable approach or procedure
2. Appropriate approach or procedure some of the time
3. Workable approach or procedure
4. Efficient or sophisticated approach or procedure

**WHY? DECISIONS
MADE ALONG
THE WAY**

*Sources of Evidence*

**a.** Changes in approach

**b.** Explanations (oral or written)

**c.** Validation of final solution

*Final Rating*

**1.** No evidence of reasoned decisions

**2.** Some evidence of reasoned decisions

**3.** Reasoned decisions are used throughout

**4.** Reasoned decisions are explicitly discussed and adjustments are considered

**WHAT? OUTCOMES
OF ACTIVITIES**

*Sources of Evidence*

**a.** Solutions

**b.** Observations and extensions (what if. . . )

**c.** Connections, applications, generalizations, syntheses

*Final Rating*

**1.** Solution without observations or extensions

**2.** Solution with observations or extensions

**3.** Solution with connections or applications

**4.** Solution with generalizations or syntheses

**LANGUAGE OF
MATHEMATICS**

*Sources of Evidence*

**a.** Terminology

**b.** Notation and symbols

*Final Rating*

**1.** No use or inappropriate use of mathematical language

**2.** Appropriate use of mathematical language some of the time

**3.** Appropriate use of mathematical language most of the time

**4.** Use of rich, precise, elegant, and appropriate mathematical language

**MATHEMATICAL
REPRESENTATIONS**

*Sources of Evidence*

**a.** Graphs, labels, charts

**b.** Models

**c.** Diagrams

**d.** Manipulatives

*Final Rating*

**1.** No use of mathematical representations

**2.** Use of mathematical representations

**3.** Accurate and appropriate use of mathematical representations

**4.** Perceptive use of mathematical representations

## CLARITY OF REPRESENTATION

*Sources of Evidence*

**a.** Audio or video tapes (or transcripts)

**b.** Written work

**c.** Teacher interviews or observations

**d.** Journal entries

**e.** Student comments

**f.** Student self-assessment

*Final Rating*

**1.** Unclear (disorganized, incomplete, insufficient detail)

**2.** Some clear parts

**3.** Mostly clear

**4.** Clear

## SCORING RUBRIC

A scoring rubric identifies and gives a value to the different levels of responses to an open-ended question. Responses are sorted into piles according to three categories: (1) demonstrates competence; (2) satisfactory response, and (3) inadequate response. Each response is then reviewed and given a point value specified within each category. Papers in the first category may receive a point value of 5 or 6. In the second category the point values are 3 or 4, and in the last category the point values are 0, 1, or 2. The criteria for each rating are given below.

### A Rating of 6

The response is complete with a clear and coherent explanation. The response includes an appropriate diagram or chart, identifies all the important elements of the problem, shows the relevant mathematical ideas and processes, and presents a strong supporting argument.

### A Rating of 5

The response is reasonably clear. It may contain an appropriate diagram or chart and show some understanding of the mathematical ideas and processes. In addition, the response identifies most of the important elements of the problem.

### A Rating of 4

The explanation has minor flaws, but is satisfactory. Some of the arguments may be incomplete or the response may be a bit unclear. The student may also have been ineffective in his/her use of diagrams or charts.

### A Rating of 3

The response has serious flaws, but is almost satisfactory. The student begins the problem correctly, but fails to complete it or may omit significant parts of the problem. The student may make flagrant computation errors, use inappropriate terminology or symbolism, or use an inappropriate strategy for solving the problem.

### A Rating of 2

The individual has started to solve the problem, but is unable to complete it. The explanation is not understandable and there is no evidence that the problem situation was completely understood.

**A Rating of 1**

The student was unable to begin the problem effectively. Parts of the problem are copied, but without working toward a solution.

**A Rating of 0**

The student made no attempt to solve the problem.

**SPECIFIC SCORING RUBRICS**

Each question or project needs to have its own rubric. The rubric needs to identify the goals and the expectations the teacher has for that particular question, so the teacher must create his/her own scoring rubric.

A scoring rubric can be created before or after reviewing some of the responses. If created before, allowances will need to be made for unusual responses. Creating the rubric after viewing a sampling of the responses may help in identifying the different levels. Either way, a scoring rubric needs to reflect what you, the teacher, values.

Creating and using a scoring rubric can be time consuming at first. To get started, work with a colleague. Working with another teacher to create the rubric and review the papers will increase your confidence in the consistency of grading. If you use the same question or project each year, save the rubric to use again, making adjustments as needed.

As your use of rubrics increases and you become more familiar with them, the amount of time spent grading holistically will decrease. Grading may still take longer than before, but the amount of information you receive from your students will also be greater.

**SAMPLE RUBRIC**

The rubric below was created for the following question.

Construct a table that shows several solutions of the equation $y = |x| + 1$. Then plot the corresponding points and describe the graphical pattern.

**A Rating of 6**

Student has constructed a table of several solutions to the equation, including $x$ values that are positive and negative. The points are plotted correctly. A description of a V-shaped graph with its point at (0,1) is clearly stated.

**A Rating of 5**

Student has constructed a table of several solutions of the equation including $x$ values that are positive and negative. The points are plotted correctly, but the description is curved instead of V-shaped.

**A Rating of 4**

The student has constructed a table of several solutions, however, only positive (or negative) $x$ values were used. The points are plotted correctly, but the graphical pattern is described as a line.

**A Rating of 3**

The response is missing a part of the problem: (1) the table is not constructed, (2) the solution points are not plotted, (3) the solution points are plotted incorrectly, (4) the graphical pattern is not described.

**A Rating of 2**

Construction of the table is started. Some points are plotted, but they do not correspond to any points from the table.

**A Rating of 1**

The outline of a table is made, but no solutions are entered.

**A Rating of 0**

Student made no attempt to do the problem.

**WHAT IS A PORTFOLIO?**

Basically, a portfolio is a collection of work completed by an individual. Professionals such as artists, writers, and models use portfolios to demonstrate their best work. In mathematics, portfolios can be used to document a student's development. When using portfolios as a means of assessment, it is important to show what the student can do, rather than what the student cannot do. A mathematical portfolio is more than just a folder of a student's work; it can be used to gain insight into the student's mathematical reasoning, understanding, ability to communicate mathematically, and attitudes.

**CONTENTS**

Work in portfolios might include any or all of the following.

1. *Open-Ended Questions, Problems, and Tasks*   The student is given an open-ended question, a problem, or a task to discuss in writing. The student is asked to formulate hypotheses, explain a mathematical situation, make generalizations, and so on.

2. *Research Projects*   The student is given a long term project that requires use of resources outside the classroom. A time line is useful to keep students on track. The time line should include dates when the following are due: list of resources, outline, rough draft, and final project.

3. *Presentations, Discussions, and Debates*   The student writes a summary of the presentation, discussion, or debate including the original assignment, the outcome, and a list of resources used in researching the topic. The names of partners should also be listed.

4. *Journal Entries*   A journal entry consists of a student's writings about mathematics. These writings could include reflections and reactions about particular assignments or class activities.

5. *Cooperative Learning Activities*   The student writes his or her own summary of the work accomplished in the cooperative learning activity. The names of others in the group should be included.

6. *Demonstrations*   Demonstrations can be done in groups, pairs, or individually, and usually involve such tools as manipulatives, graph paper, compasses, calculators, or computers.

7. *Math Logs*   Math logs are worksheets that contain writing activities correlated to the lessons.

8. *Investigations*   The student keeps a log that includes the date, a description of the work done, and any questions to the teacher the student may have. The teacher's response to the question is written in the log.

9. *Photographs*   Items that are too bulky to fit into a portfolio can be photographed and the photograph included in the portfolio.

10. *Models and Simulations*   The student writes a summary that includes the original assignment and an explanation of the model or simulation. Diagrams, sketches and/or photographs should be included.

11. *Problem Solving*   A portfolio can contain the solutions of non-routine problems solved using the following five steps.

   • Read the problem. Be able to restate it in your own words.

   • Explore the problem, draw a picture, make charts, make diagrams.

   • Choose a strategy such as guess and test, look for a pattern, logical deduction, reduction, simulation, working backwards, and exhaustive listing.

   • Carry out the strategy and solve the problem.

   • Look back.  Ask yourself, Is there another strategy I could have used to solve the problem? Is this a unique solution? Can I make a generalization? What if . . . (extend the problem)?

12. *Interviews*   During an interview, students talk about a problem they are solving while the teacher listens and asks questions. Students may be interviewed individually or in groups. Some examples of phrases used to encourage students to further elaborate on their explanations are as follows.

   • I am interested in your thinking.

   • Please help me to understand. Let's suppose that you are the teacher, and I am your student.

   • Sometimes when I am having trouble with a problem, I break it down into small steps. Let's try to do that now.

   • I understand it better now, but . . .

   • I like it when you take time to think about the problem and your explanation.

   Notes taken from the interview by the teacher or another student of the group may be included in the portfolio.

13. *Time-Staggered Sampling*   The portfolio can contain the results of work dealing with the same mathematical idea sampled at different times.

14. *Awards and Prizes*   The portfolio can contain descriptions or copies of awards and prizes that the student has won.

**SPECIAL FEATURES**

The following special features should be included in each student's portfolio: (1) table of contents, (2) identification of who selected the piece of work, (3) dates on all work, (4) description of the problem or task, (5) cover letter, and (6) comment sheet including self-assessment.

**SELECTION PROCESS**

Both the teacher and the student should have input into the selection process. The teacher may decide how many pieces are to be included, and the categories from which these pieces are to come. The student would then be allowed to choose the pieces, and have a comment sheet explaining how and why each of the pieces was chosen.

**WHAT IS AN
ERROR NOTEBOOK?**

An error notebook gives students an opportunity to analyze and to learn from their errors.

To make an error notebook, follow the steps below.

1. Divide a notebook page into three columns as shown.

2. The first column should contain the problem on which the errors were made. The source (homework, quiz, or test) of the problem should also be included.

3. The second column should display exactly the error that was made. Draw a sad face next to the error to highlight the error that was made. The student might also include a statement such as "I drew a blank on this one," as an explanation of what went wrong. You might wish your students to use red ink for this column since this is the real source of the student's learning experience.

4. The third column should contain comments. The corrected problem should be displayed, including any thoughts or concepts that pertain to the problem.

5. Separate the problems with horizontal lines. Do not restrict the length of any column. Allow the students to write as much as they feel is necessary.

6. Students should spend a few minutes each day reading the notebook.

**SAMPLE FORM**

| NAME_____ | | ERROR NOTEBOOK |
|---|---|---|
| **Problem** | **Error** | **Correction & Comment** |
| 1. (Quiz on Oct. 20) Write an equation that represents the sentence and solve the equation. The product of the price of a can of soda, $p$, and 5 cans is $3.00. | $5 \cdot 3 = p$ $\$15 = p$ <br> ☹ | $5p = 3$ $p = 0.6$ <br> The product is the result of multiplication. Thus, the product of $p$ and 5 is $5p$. The word "is" indicated equality. Thus, the product $5p$ equals 3. |
| 2. | | |

**WHAT IS A MATH JOURNAL?**

A math journal can be used by students to assess their own progress in the course and to assess their attitude about the course.

You can ask students to complete journal entries on a regular basis (daily or weekly), or on an occasional basis. The questions used for the journal are up to you. The following sample is one that has been used by teachers on a daily basis.

**SAMPLE FORM**

DATE _____ M T W R F                                        Math Journal

1. What were the goals of today's math lesson?
   _____
   _____
   _____
   _____

2. Why did I learn this?
   _____
   _____
   _____

3. What strategies can I use to accomplish today's goals?
   _____
   _____
   _____
   _____

4. What did I like best about today's math class?
   _____
   _____
   _____
   _____

5. What was most frustrating about today's class?
   _____
   _____
   _____
   _____

**WHAT IS A NOTEBOOK QUIZ?**

A notebook quiz is a one-page quiz that can be used as a substitute for collecting and grading notebooks. This gives students constant access to their notebooks while giving the teacher feedback on the completeness of the students' notebooks. Grading can be done over a period of a few days because the students retain possession of their notebooks.

If used, notebook quizzes should be given in class regularly. Students may use their notebooks to answer the questions, but may not use their textbooks or any other sources. Students should be told at the beginning of the course what type of information they need to keep in their notebooks. For instance, they should record dates, goals, definitions, and so on. Students also need to know that they are responsible for obtaining notes that were missed due to absence.

Here is a sample of the type of questions you might consider giving on a notebook quiz.

**SAMPLE FORM**

PASSPORT to ALGEBRA and GEOMETRY      NAME _____
NOTEBOOK QUIZ      DATE _____

1. What is the name of the chapter we have been studying?

2. State the goals discussed in class on January 27.
   Goal 1:
   Goal 2:

3. Define the term congruent.

4. State one of the goals discussed in class on January 31.

5. State the formula used to find the measure of an interior angle of a regular polygon.

6. State one of the goals discussed in class on February 2.

7. Complete the statement. In a triangle, the longest side is opposite

8. What is the relationship between the sides and angles of an isosceles triangle?

**WHAT IS A DAILY QUIZ?**

The teacher can use the daily quiz to determine if students need additional help or are ready for enrichment activities. If the class has a firm understanding of a topic, then less classroom time is needed for review and can instead be used in developing new concepts.

A daily quiz also provides the teacher with opportunity to assess the students' understanding of the previous day's lesson and homework assignment. The students are encouraged to prepare for the quiz by paying attention in class and completing the homework assignment.

**THE DAILY QUIZ**

Four problems should be on the chalkboard or the overhead so that the students can begin the quiz as soon as they enter the classroom. Students can grade each other's papers and keep them in their notebooks to be turned in on Friday, or the teacher can collect and distribute the papers daily. If each problem is worth 5 points, then the daily quizzes total 100 points by the end of the week.

**SAMPLE FORM**

An $8\frac{1}{2}$ by 11 sheet of paper can be divided into five rows and five columns. The first column lists the days of the week. The remaining four columns are for the four problems given each day.

| Name _____ | | | | |
| Class _____ | | | | |
|---|---|---|---|---|
| Monday | | | | |
| Tuesday | | | | |
| Wednesday | | | | |
| Thursday | | | | |
| Friday | | | | |

**WHAT IS A PROJECT?**

A project is an undertaking that requires effort over time. In this booklet, projects are divided into many headings: Research Project, Journal Entry, Construction, Demonstration, Interview Assessment, Problem Solving, Math Game, Cooperative Learning, Open-ended Question, and Discussion/Journal Entry. The projects can be used to extend the lesson, to provide an opportunity to observe your students, or to give students added insight into the algebra and geometry of the world.

Any of the projects can and should be adapted to your classroom. A construction problem could become a demonstration done by a group of students for the rest of the class. An open-ended question could lead to a group discussion. Students can make a journal entry to assess their attitudes and feelings about a completed project.

**GRADING**

Students should be aware of the grading criteria of a project when the project is assigned. If a scoring sheet is to be used, pass out a copy to the students so they are aware of the value of each portion of the assignment. Once graded, a project can be a wonderful addition to a student's portfolio.

| COOPERATIVE LEARNING |
|---|

*(Use after Lesson 1.1)*

1. Divide the class into groups of four.

2. Give each group a road map.

3. Have each group find as many different uses of numbers on the map as possible.

4. Have each group categorize each use as an identifier or a measurement.

5. Have each group find any sequences.

**Assessment Goal:**

- Evaluate student's knowledge of how numbers and sequences are used to identify and measure real-world objects.

| COOPERATIVE LEARNING |
|---|

*(Use after Lesson 1.3)*

1. Divide the class into groups of five.

2. Give each member of a group a note card with the following exercises:

   NOTE CARD 1: $\sqrt{17}, \sqrt{5}, \sqrt{10}, \sqrt{34}$

   NOTE CARD 2: $\sqrt{3.8}, \sqrt{2.1}, \sqrt{7.3}, \sqrt{11.2}$

   NOTE CARD 3: $\sqrt{0.5}, \sqrt{0.3}, \sqrt{0.9}, \sqrt{0.02}$

   NOTE CARD 4: $\sqrt{\frac{1}{2}}, \sqrt{\frac{5}{9}}, \sqrt{\frac{3}{4}}, \sqrt{\frac{2}{3}}$

   NOTE CARD 5: $\sqrt{\frac{10}{3}}, \sqrt{\frac{8}{7}}, \sqrt{\frac{21}{4}}, \sqrt{\frac{101}{100}}$

3. Have each member of the group find the values of the expressions on their card using a calculator.

4. Have each member of the group decide whether the square root or the original number is larger.

5. Have the group compare answers.

6. Have the group explain the conflicting results.

**Assessment Goal:**

- Discover the effect of the square root operation.

Insert parentheses to make the number sentences true.

**a.** $16 \div 4 + 4 + 3 = 5$    **b.** $12 \div 3 \cdot 2 + 7 = 9$

**c.** $2 + 4 \div 2 \cdot 5 = 15$    **d.** $2 + 3^2 - 6 = 19$

**Assessment Goal:** • Develop proper use of order of operations.

ORAL PRESENTATION (Use after Lesson 1.6)

**1.** Divide the class into groups of four.

**2.** Let each group select a research topic. Some suggestions for topics are as follows:

   **a.** Number of Republicans and Democrats in the U.S. Senate from 1980–1990.

   **b.** Number of gold, silver, and bronze medals won by the U.S. in the past 10 Winter Olympic Games.

   **c.** Measure the temperature of 8 ounces of tap water as it heats on a stove at a medium setting. Starting at 0 seconds, collect data every 30 seconds for 5 minutes.

   **d.** Dow Jones Industrial Average for the past 10 business days.

   **e.** Number of boys and girls that have graduated from your school over the past 10 years.

   **f.** The number of different number one songs on the country and pop charts for each month of the past year.

   These topics may be altered. Another option is to let the group pick their own topic.

**3.** Have the group decide upon one member for each of the following positions:

   **a.** Researcher. The researcher collects all the data to be used as well as any other information that will be presented as part of the oral report.

   **b.** Graphic Designer. The graphic designer is in charge of the selection of the type of graph to be used and its construction.

   **c.** Writer. The writer produces the general manuscript for the presentation. This includes incorporating any background information with graph analysis.

   **d.** Presenter. The presenter gives the oral presentation to the entire class and fields any questions from the audience.

**4.** Although each member of the group has certain responsibilities, interaction among group members should be strongly encouraged.

**Assessment Goals:**

• Evaluate students' ability to work in groups.

• Evaluate students' ability to prepare and make a presentation.

• Evaluate students' understanding of the use of graphs in the organization of data.

## Investigations in Algebra

**COOPERATIVE LEARNING**   *(Use after Lesson 2.2)*

1. Divide the class into groups of four.

2. Give each group one of the following expressions:

    **a.** $0.5x + 6$        **f.** $7x + 3y$

    **b.** $10x + 25$        **g.** $3xy$

    **c.** $3.50x$           **h.** $x^2$

    **d.** $30 - 2.75x$      **i.** $a + b + c$

    **e.** $3x - 100$        **j.** $6.60x$

3. Have each group make up a real world problem that could be modeled by their expression.

    EXAMPLE:  How much would it cost to buy three books of stamps?

4. Collect the groups' exercises and make a worksheet with all of the questions.

5. Using the same groups, have each group complete the worksheet.

**Assessment Goal:**

- Evaluate students' understanding of the connection between real-world problems and algebra.

**EXPERIMENTATION**   *(Use after Lesson 2.4)*

1. Divide the class into groups of four.

2. Provide each group with a two-plate balance, two 5 gram weights, four 1 gram weights, a stack of pennies, a collection of paper clips, several pieces of whole chalk, several new unsharpened pencils, and a stack of notebook paper. (You may substitute or add a variety of other items.)

3. Have the groups place any combination of known weights on one plate. Have the groups place enough pennies on the other plate to make the scale balance.

4. Let $x$ be the weight of one penny. Have the groups write an equation that represents the condition of the two-plate balance.

5. Have the groups determine the weight of a single penny by solving the equation.

6. Have the groups repeat this experiment and estimate the weight of a single paper clip, piece of chalk, pencil, and piece of notebook paper.

**Assessment Goals:**

• Develop equation solving skills.

• Demonstrate a physical meaning of equations.

---

**COOPERATIVE LEARNING**

*(Use after Lesson 2.6)*

1. Divide the class into pairs.

2. Give each pair the following recipe. (Any recipe with a variety of measures and a standard number of servings can be used.)

Broccoli Soup

| | |
|---|---|
| $1\frac{1}{2}$ lbs | broccoli |
| $\frac{1}{2}$ cup | onion |
| 2 Tbsp | margarine |
| 2 Tbsp | flour |
| $4\frac{1}{2}$ cups | water |
| 1 Tbsp | chicken bouillon |
| $\frac{3}{4}$ tsp | salt |
| $\frac{1}{8}$ tsp | pepper |
| $\frac{1}{2}$ cup | cream |

This recipe makes 8 servings.

3. Have the pairs answer the following questions.

   **a.** You need to make enough soup for 6 servings. Write an algebraic model that gives the amount of an ingredient in terms of the given quantity.

   **b.** Determine the amount of each ingredient needed.

   **c.** If the only measuring supplies you have are $\frac{1}{2}$ cup, $\frac{1}{8}$ cup, 1 tsp., $\frac{1}{2}$ tsp., and $\frac{1}{4}$ tsp., which of the ingredients can you accurately measure?

**Assessment Goals:**

• Development of students' modeling ability.

• Development of appreciation of real-world applications.

**COOPERATIVE LEARNING**   *(Use after Lesson 3.4)*

1. Divide the class into groups of five.

2. One member of the group plays the part of the mail carrier. The other members of the group receive mail. The role of the mail carrier should be changed occasionally during the activity.

3. The mail carrier may do one or more of the following:

   **a.** deliver a bill            **b.** deliver a check

   **c.** take away a bill that      **d.** take away a check that
   was delivered in error           was delivered in error

4. Give the mail carrier a set of instructions as to what should be delivered to each member of the group. Provide enough information so that multiple deliveries are made. Initially, each member of the group starts with a balance of $20.

5. After the mail is delivered, each group member must decide how much money they have. Also have the members describe the integer operation they used to arrive at their conclusion.

   EXAMPLE:  A student receives a bill for $10, a check for $5, and is told that a previously delivered bill for $2 is being taken back. If the student had $20 prior to the mail delivery, he would use the calculation, $20 - 10 + 5 - (-2)$, for a total of $17.

**Assessment Goal:**

- Development of students' ability to use integer operations.

**PROBLEM SOLVING**   *(Use after Lesson 3.6)*

Have the students solve the given puzzles. Students may work individually or in groups.

1. Given $ab < 0$, $a|c| > 0$, $|d| = d$, and $\dfrac{d}{c} > 0$, what is the sign of $(a + c)b$?

2. Given $a - c < 0$, $ac < 0$, $cd > 0$, and $\dfrac{b}{d} < 0$, what is the sign of $\dfrac{a + b}{d} - c$?

3. Given $a - b > 0$, $ab > 0$, $abc > 0$, $cd > 0$, and $\dfrac{b}{d} < 0$, what is the sign of $\dfrac{a - c}{b - d}$?

4. Given $a - b > 0$, $c - d < 0$, $cd < 0$, and $abcd > 0$, what is the sign of $c\left(\dfrac{d}{a - b}\right)$?

**Assessment Goal:**

- Development of logical thinking.

## JOURNAL ENTRY

*(Use after Lesson 3.7)*

The computer club wants to buy a new printer that costs $300. They already have $50 in a bank account. To earn the extra money, they decide to hold a car wash. The only cost to the club is the price of soap. A local company will supply soap so that they will only pay for what they use. Decide which of the following verbal models accurately represent the situation. Write a paragraph explaining why the model(s) are correct representations. Also, describe what is wrong with the incorrect model(s).

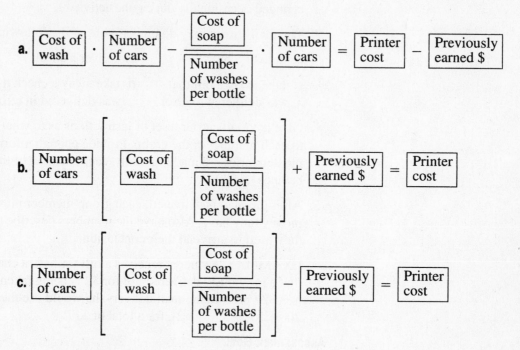

**Assessment Goals:**

- Development of modeling skills.
- Development of writing skills.

## COOPERATIVE LEARNING    *(Use after Lesson 3.8)*

1. Divide the class into groups of three. Have each group suspend a board so that it is 1 foot off the ground. (A broom works well.)

2. Have each group release a tennis ball (or any ball that will bounce) 2 feet off the ground. Count and record the number of times the ball bounces before it reaches the 1 foot level.

3. Repeat this process when the ball is released at different heights.

4. Have each group make a table and plot their data.

5. Have each group describe any patterns that arise.

**Assessment Goals:**

- Develop data organization skills.
- Develop graph analysis skills.

**PROBLEM SOLVING** *(Use after Lesson 4.5)*

1. Divide the class into groups of four.

2. Give each group the following exercises.

   a. $-2(x + 7) + 2 + x(3 + 4) = -3(x - 2) + 2x + 4 - 5x$

   b. $4x + 2(x - 1) + 3(2x + 5) = 4x - 3x + 2(2x + 1) - 8$

   c. $2 + 3(2 + 4x) - 5x + 2(x - 1) = 5(4x - 3) - 8x$

   d. $2x - 6x + 4(2x - 3) + 7 = -2(4 + x) + 2(3x + 1) - 5x + 3$

3. Have each member perform one operation on the exercise and pass the exercise to the next group member. Continue this process until the equation has been solved.

4. Have the group check their solution and find any errors.

**Assessment Goal:**

- Develop equation solving skills.

**COOPERATIVE LEARNING** *(Use after Lesson 4.6)*

1. Divide the class into groups of four.

2. Provide each group with two pieces of tracing paper on which is drawn a coordinate plane and two different colored pens or pencils.

3. Given the table below, have two of the group members plot the points (year, farm occupation). Have the other students plot the points (year, nonfarm occupation). Each pair should use a different color. *(Source: Information Please Almanac, Atlas, and Yearbook, 1997)*

| Year | 1840 | 1850 | 1860 | 1870 | 1880 | 1890 | 1900 | 1910 |
|---|---|---|---|---|---|---|---|---|
| Percent of Labor Force in Farm Occupation | 68.6 | 63.7 | 58.9 | 53.0 | 49.4 | 42.6 | 37.5 | 31.0 |
| Percent of Labor Force in Nonfarm Occupation | 31.4 | 36.3 | 41.1 | 47.0 | 50.6 | 57.4 | 62.5 | 69.0 |

4. Have each pair connect their points. Then have the pairs come together and overlay their graphs.

5. Have the students answer the following questions:

   a. In what year was the percentage of workers in a farming occupation the highest?

   b. In what year was the percentage of workers in a nonfarming occupation the highest?

**c.** In what year were the percentages about the same?

**d.** In what year was the difference in percentage the greatest?

**e.** In what year(s) was the percentage of jobs in farming occupations greater than the percentage of jobs in nonfarming occupations?

**Assessment Goals:**

- Develop graph construction skills.

- Develop students' ability to interpret graphs.

---

**COOPERATIVE LEARNING**   *(Use after Lesson 4.7)*

**1.** Divide the class into groups of four.

**2.** Give each group the following definitions and example.

Definition:  A procedure that is repeated over and over again and in which each step uses the results obtained from the previous step is called an iterative process. Each application of the procedure is called an iteration.

Example:  Repeat the following procedure until the result is greater than 10. Start with 1 and double the given number.

Initial step:   1
Iteration 1:   2
Iteration 2:   4
Iteration 3:   8
Iteration 4:   16  STOP.

**3.** Give the group the following iterative process. Start with 0.9. Use a calculator to square the given number. One member of the group should round the results after each iteration to one decimal place. The other members should round the results after each iteration to 2, 3, and 4 decimal places, respectively. Repeat this process until the calculator displays 0.

**4.** Have the group members compare their results and answer the following questions.

**a.** How many iterations did you have?

**b.** How did each iteration compare?

**c.** Does round off error affect the iterative procedure? How?

**5.** Another iterative process to try: Start with $\sqrt{2}$, take the square root of the number, and stop when you reach 1. Round the results after each iteration to 1, 2, 3, and 4 decimal places, respectively.

**Assessment Goal:**

- Develop appreciation for the effect of round off error.

## Suggested Topics

**a.** USA Space Program

**b.** Development of the Computer

**c.** Revolutionary War

**d.** Lifetime of Shakespeare

**e.** Music Compositions of Mozart

**f.** Lifetime of Martha Graham

**g.** Automobiles

**h.** Development of the Polio Vaccine

## Suggested Questions

**a.** How long have you been teaching?

**b.** How long have you been teaching at this school?

**c.** What was your favorite subject in school?

**d.** Did you participate in any extra-curricular activities when you were in junior high or high school? If so, which ones?

**e.** In what state were you born?

**f.** What grade (or subject) do you teach?

**g.** Who were your role models as a student?

### RESEARCH PROJECT  *(Use after Lesson 5.1)*

1. Divide the class into groups of three.

2. Have each group research a topic.

3. Have each group draw a time line that includes important data found in their research.

4. An oral presentation could also be incorporated into this project.

**Assessment Goal:**

- Develop data organization skills.

### JOURNAL ENTRY  *(Use after Lesson 5.2)*

Find an example of a bar graph in a newspaper or magazine that you may cut out. Tape the picture in your journal and write a paragraph that analyzes the graph.

**Assessment Goals:**

- Develop students' ability to analyze graphs.

- Develop writing skills.

- Develop students' ability to apply mathematics to the real world.

### INTERVIEW  *(Use after Lesson 5.2)*

1. Obtain authorization from other teachers at your school to have your students interview them.

2. Divide the class into pairs. Give each pair the name of one or more teachers along with their classroom location.

3. Give each pair a list of interview questions. As an alternative, have the class as a whole choose interview questions. Impress upon your class the importance of choosing questions that will not invade the privacy of the teachers. Also make sure that the students understand that any teacher may refuse to answer one or more of the questions.

4. Collect all of the data and have the class construct a histogram or bar graph to display the data for each question.

5. Finally, have a class discussion about the results.

**Assessment Goal:**

- Develop students' ability to collect and organize data.

**OPEN-ENDED QUESTION**  *(Use after Lesson 5.7)*

1. Place a long list of items (groceries, names, places, mathematicians, dates, etc.) on the overhead. The lists should contain about 50 items.

2. Allow the students to study the list for 10 seconds.

3. Have the students write down as many items on the list as they can recall. Have them count up and record the number of items they remembered.

4. Allow the students to study the list for an additional 10 seconds. Again, have them try to recall as many items on the list as possible and record the number.

5. Repeat this process as least ten times.

6. Have the students create a scatter plot made up of data points of the form (attempt, items recalled). For example, the data point (3, 10) indicates that on the third attempt, 10 items were remembered. This plot is their "learning curve."

7. Have the students determine if there is a correlation between the data. If so, what is the correlation?

**Assessment Goals:**

- Develop graphing skills.

- Develop students' ability to find correlations.

**EXPERIMENT**  *(Use after Lesson 5.8)*

1. Divide the class into pairs.

2. Assign each pair a topic to research at the library. Some examples of topics are given below:

   Number of households in the United States that have a pet (dog, cat, bird, turtle, or rabbit)

   Blood types

   Number of phones per household in the United States

   Average number of hours of sleep per night

3. Have each pair design and conduct a poll on their topic using the entire class as the polled population.

4. Have the pairs compare the results of their in-class poll with the result discovered at the library. If appropriate, let the pairs present their findings to the class.

5. Be sure to have the pairs discuss the following questions:

   **a.** Did the data collected from the class match the results of your library research?

   **b.** Explain why differences may have occurred.

   **c.** Other than the size of the sample group, can you think of other reasons why data collected in a poll may not accurately reflect the facts?

**Assessment Goal:**

- Develop students' understanding of polls and probability.

**MATH GAME** *(Use after Lesson 6.1)*

1. Divide the class into groups of four. Have each group sit in a circle.

2. The group counts. The first student begins with one, the second student responds with two, and so on. All numbers that have seven as a factor or a digit must be skipped. Example: After a student calls 13, the next student calls 15 (skipping 14, because 7 is a factor).

3. If a student gives an incorrect answer, they must sit out of the game. Any eliminated students continue to monitor the answers given by the remaining players.

4. The game is continued until one student remains.

5. This game can be repeated using any other number from 2 to nine as the "skipped" number.

**Assessment Goal:**

- Evaluate students' ability to recognize factors of natural numbers.

**MATH GAME** *(Use after Lesson 6.5)*

1. Divide the class into groups of six.

2. Give each group a "deck" of note cards with the following fractions written on them.

| | | | | | | | |
|---|---|---|---|---|---|---|---|
| $\frac{1}{6}$ | $\frac{3}{18}$ | $\frac{5}{30}$ | $\frac{9}{54}$ | $\frac{7}{8}$ | $\frac{14}{16}$ | $\frac{21}{24}$ | $\frac{42}{48}$ |
| $\frac{2}{7}$ | $\frac{6}{21}$ | $\frac{10}{35}$ | $\frac{16}{56}$ | $\frac{8}{9}$ | $\frac{16}{18}$ | $\frac{32}{36}$ | $\frac{72}{81}$ |
| $\frac{1}{3}$ | $\frac{4}{12}$ | $\frac{6}{18}$ | $\frac{9}{27}$ | $\frac{6}{5}$ | $\frac{24}{20}$ | $\frac{30}{25}$ | $\frac{42}{35}$ |
| $\frac{3}{7}$ | $\frac{9}{21}$ | $\frac{21}{49}$ | $\frac{27}{63}$ | $\frac{5}{3}$ | $\frac{20}{12}$ | $\frac{30}{18}$ | $\frac{50}{30}$ |
| $\frac{1}{2}$ | $\frac{3}{6}$ | $\frac{6}{12}$ | $\frac{9}{18}$ | $\frac{3}{2}$ | $\frac{15}{10}$ | $\frac{21}{14}$ | $\frac{24}{16}$ |
| $\frac{2}{3}$ | $\frac{4}{6}$ | $\frac{10}{15}$ | $\frac{12}{18}$ | $\frac{9}{4}$ | $\frac{27}{12}$ | $\frac{54}{24}$ | $\frac{72}{32}$ |
| $\frac{4}{5}$ | $\frac{8}{10}$ | $\frac{24}{30}$ | $\frac{32}{40}$ | | | | |

3. Three students will play the card game and three will be referees. Each of the players will have a referee watching his or her hand.

4. The game is played like "Go Fish." The goal of the game is to earn the most points by collecting pairs of equivalent fractions.

   a. The cards should be well shuffled.

   b. Each player is dealt 7 cards.

   c. The players lay down any pair of cards that they are holding which have equivalent fractions written on them.

**d.** If player 1 has an $\frac{8}{10}$ card, he or she may ask any one player if he or she has an $\frac{8}{10}$. If the asked player has a card with an equivalent fraction written on it, he or she must give it to player 1. Player 1 must now lay the pair down. Player 1 then asks any one player about another card. If the asked player does not have an equivalent fraction, player 1 must draw a card from the pile. The next player begins.

**e.** If players say that they do not have an equivalent fraction when they do, the appropriate referee stops play and corrects the error.

**f.** The game is over when one player runs out of cards.

**g.** Each pair is worth 1 point. For each time the referee had to correct you, one point is subtracted from your score.

**5.** Have the groups play again after switching the roles of players and referees.

**Assessment Goal:**

• Develop students' ability to compare fractions.

---

| COOPERATIVE LEARNING | *(Use after Lesson 6.9)*

**1.** Divide the class into groups of four.

**2.** Give the groups the following definition.

**Definition:** A geometric sequence is a list of numbers in which any number, after the first, can be obtained by multiplying the previous number by a fixed number, called a common ratio.

**3.** Give the groups the sequence 1, 2, 4, 8, . . . , and have them answer the following questions.

**a.** What is the first number in the sequence?

**b.** What is the common ratio?

**c.** What is the 6th number?

**d.** Find a pattern and write a formula for the $n$th number in the sequence.

**e.** What is the 20th number?

**4.** Have the group repeat step 3 for the following geometric sequences.

**a.** 3, 6, 12, 24, . . .          **b.** 4, 2, 1, $\frac{1}{2}$, . . .

**5.** Have the groups discuss and answer the following questions.

**a.** Among all three examples, is there a pattern in the formulas for the $n$th number? If so, what is this pattern?

**b.** Without looking at the specific numbers, write a formula for the $n$th number of a geometric sequence with a common ratio of $\frac{2}{3}$ and 5 as the first number.

**Assessment Goals:**

• Evaluate students' ability to recognize patterns.

• Evaluate students' ability to work with powers.

### JOURNAL ENTRY  *(Use after Lesson 7.2)*

When adding or subtracting unlike fractions, it is not necessary to use the least common denominator. Any common denominator will work. Explain why, and demonstrate this conclusion with an example. Discuss any advantages or disadvantages to using the *least* common denominator.

**Assessment Goals:**

- Develop writing skills.

- Develop students' ability to solve a problem in more than one way and to determine the efficiency of the methods.

### JOURNAL ENTRY  *(Use after Lesson 7.6)*

"You must give 110%!" is a common saying when referring to the effort required to succeed in athletics, business, school, etc. Is it actually possible to give 110% of yourself to a project? Write a paragraph explaining your answer. Also include the meaning of the saying.

**Assessment Goals:**

- Develop students' understanding of percents.

- Develop writing skills.

### COOPERATIVE LEARNING  *(Use after Lesson 7.8)*

1. Divide the class into groups of four.

2. Have the groups discuss the given situation and answer the questions.

   \* Many casinos advertise in bright lights that there is a 98% return on money gambled in their establishments.

   **a.** Does this sound like there is a good chance of winning?

   **b.** Complete the table.

| Money Bet | Money Returned at 98% of the Bet | Profit/Loss |
|-----------|----------------------------------|-------------|
| $10 | | |
| $20 | | |
| $30 | | |
| $40 | | |
| $50 | | |
| $100 | | |
| $1000 | | |

**c.** If you make the 98% return, do you win or lose money?

**d.** What type of return would you need to actually profit?

**e.** Why do you think casinos use this advertising technique?

**Assessment Goals:**

- Develop students' ability to interpret percentages.

- Develop consumer awareness.

---

| COOPERATIVE LEARNING |  *(Use after Lesson 7.9)*

**1.** Divide the class into groups of four.

**2.** Give the groups the following information.

   \* You deposit $200 in a savings account that pays 4.25% in simple interest. During the year, you make no deposits or withdrawals. At the end of each year, you deposit an extra $100.

**3.** Have the groups discuss and answer the following questions.

   **a.** Complete the table.

| Year | Initial Balance | Interest Earned | Ending Balance |
|------|-----------------|-----------------|----------------|
| 1    |                 |                 |                |
| 2    |                 |                 |                |
| 3    |                 |                 |                |
| 4    |                 |                 |                |

   **b.** If the above scenario was altered by making one of the following changes, which change would be more profitable? (You may wish to make other tables.)

   I. Find an account that pays 4.5% in simple interest.

   II. Deposit an extra $105 at the end of each year.

   III. Find an account that pays 6.75% in simple interest.

   **c.** Complete the following statements with *always, sometimes,* or *never.*

   \*In general, increasing the initial deposit [?] increases the final balance.

   \*In general, increasing the interest rate [?] increases the final balance.

   \*In general, increasing the amount of the additional deposits [?] increases the final balance.

   \*In general, increasing the interest rate [?] has more effect on the balance than increasing the amounts of the deposits.

**Assessment Goals:**

- Develop students' understanding of simple interest.

- Develop skills in using percentages.

| JOURNAL ENTRY | *(Use after Lesson 8.1)* |

Look at the sports page (or section) of any newspaper. Find examples of both ratios and rates. Write a short statement about what information the ratios and rates give the reader.

**Assessment Goal:**

- Demonstrate real-life applications of ratios and rates.

| DEMONSTRATION | *(Use after Lesson 8.2)* |

1. Have each student find their pulse either on their neck or wrist.

2. Have the students count the number of beats in 10 seconds.

3. Have the students estimate the number of beats per minute.

4. Have the students actually count the number of beats in one minute. Compare these results with the estimate.

5. Have a class discussion on why these numbers may not be exactly the same.

**Assessment Goal:**

- Develop skills using proportions.

| COOPERATIVE LEARNING | *(Use after Lesson 8.3)* |

1. Divide the class into groups of five.

2. Provide each group with a tape measure.

3. Using the following procedure, have the groups approximate the number of people that could stand on a football field.

   a. Take a guess.

   b. Look up the size of a football field in a reference book and find its area.

   c. Estimate the area the group members occupy when standing.

   d. Write a verbal model that describes the relationship between the number of people on a football field, the number of people in your group, the area of the field, and the area needed for your group.

   e. Write an algebraic model and solve.

   f. Compare the result with your initial guess.

4. This project may be expanded by then asking how many people could stand in your classroom, the gymnasium, or your home state.

**Assessment Goals:**

- Develop estimation skills.

- Develop skills using proportions.

**RESEARCH PROJECT**   *(Use after Lesson 8.5)*

1. Divide the class into groups of four.

2. Assign each group a type of food.

   EXAMPLES: milk          frozen vegetables

   cookies          salad dressing

   soda            spaghetti sauce

   fruit juice        yogurt

3. Each group must select four different products from their category. Their selections may be different types of one brand or four different brands.

4. Each gram of fat contains 9 calories. Each gram of protein contains 4 calories. Have each group determine the percent of calories that come from fat and protein for one serving of each of their products.

5. Explain why the percent found in Exercise 4 is different from the percent listed on the label.

6. Have each group prepare a short report which discusses their findings.

**Assessment Goals:**

• Develop skills using percents.

• Encourage consumer awareness.

**EXPERIMENT**   *(Use after Lesson 8.8)*

1. Have the class determine the theoretical probability of tossing a coin and having the result "tails."

2. Toss a coin 10 times. Record the results. Ask the class how the ratio of "tails" to tosses in the experiment compared to the theoretical probability.

3. Give each student a coin. Have each student toss the coin 10 times and record their results.

4. Collect data from the first row only and have the class compare the results to the theoretical probability.

5. Repeat this process using data from the first three rows of students.

6. Finally, include the entire class' data and have them compare the results to the theoretical probability.

7. Ask the class to theorize about the connection between theoretical and experimental probabilities as a larger amount of data is used in the experiment.

8. This project may be altered by comparing theoretical and experimental probabilities of rolling a 5 on a die.

**Assessment Goal:**

• Develop an understanding of the connection between theoretical and experimental probabilities.

**COOPERATIVE LEARNING** *(Use after Lesson 9.2)*

1. Divide the class into groups of four.

2. Give each group a piece of paper with a real number line including integers from −5 to 5.

3. Have each group member write eight real numbers on individual slips of paper. The numbers should be between −5 and 5. Each member should include 1 natural number, 2 integers, 3 rational numbers, and 2 irrational numbers.

4. Put the slips of paper in a box.

5. Each member of the group takes a turn drawing a number and plotting it on the number line. The student must also state whether the number is a natural number, an integer, a rational number, or an irrational number. If a number is repeated, redraw.

**Assessment Goal:**

- Develop students' ability to compare different types of numbers.

**CONSTRUCTION** *(Use after Lesson 9.3)*

1. Divide the class into groups of four.

2. Provide each group with a ruler, a protractor, and a piece of paper with a right triangle with 1 in. legs drawn about one third of the way down the page.

3. Instruct the first member of the group to calculate the length of the hypotenuse.

4. The second group member creates a new right triangle by drawing a leg of length 1 in. and using the old triangle's hypotenuse as the new triangle's second leg. Then have the student calculate the second triangle's hypotenuse using Pythagorean's Theorem.

5. Step 4 is repeated until each group member has had the opportunity to construct three triangles. Students should draw the triangles so the 1-inch legs are adjacent.

6. Have the group measure the hypotenuse of each triangle and find a decimal approximation of the length found using the Pythagorean Theorem. How do these numbers compare?

7. The shape of the figure constructed is called a root spiral. Have the group go to the library and research what type of sea shell has this shape.

**Assessment Goal:**

- Develop students' ability to use the Pythagorean Theorem.

| COOPERATIVE LEARNING | (Use after Lesson 9.4)

1. Divide the class into pairs.

2. Assign each pair one of the fifty states.

3. Have each pair go to the library and find a map of their state in an atlas.

4. Have each pair find any three cities in the state that may act as vertices of a right triangle.

5. Each pair should use a ruler and the map's key to determine the distance between the cities that make up the legs of the right triangle.

6. Have the pairs use the Pythagorean Theorem to find the distance between the two cities that make up the hypotenuse of the right triangle.

7. Have the pair also find this distance using a ruler and the map's key and compare the results.

**Assessment Goals:**

• Develop students' understanding of map scales.

• Develop familiarity with the Pythagorean Theorem.

| COOPERATIVE LEARNING | (Use after Lesson 9.7)

1. Divide the class into groups of four.

2. Wrap stacks of 3 pennies in paper. Use paper and tape that are light weight so that the weight of the wrapper is negligible. Make enough stacks so that each group has 8 stacks.

3. Give each group a two-plate balance, 8 stacks, and 8 single pennies.

4. Have the group make sure the scale is balanced when nothing is on either plate.

5. Have the groups place 2 stacks plus 2 pennies on the left plate. Place three stacks plus one penny on the right plate.

6. Let $s$ represent the number of pennies in each stack. Have the group write an inequality that describes the situation.

7. Have the group solve the inequality. During each step, have the group model what is happening algebraically with the balance.

8. Have the group determine what this solution tells them about the number of pennies in the stack.

9. Repeat this process for the following initial conditions:

| Left Plate | Right Plate |
|---|---|
| a. 4 stacks + 3 pennies | 2 stacks + 5 pennies |
| b. 1 stack + 6 pennies | 4 stacks |
| c. 3 stacks + 2 pennies | 2 stacks + 5 pennies |

**Assessment Goal:**

• Develop an understanding of inequalities.

**COOPERATIVE LEARNING** (Use after Lesson 10.1)

1. Divide the class into pairs.

2. Provide each pair with a ruler and two pieces of paper with a right triangle drawn one third down each page. The triangle should be labeled as shown at the right.

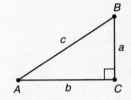

3. Have each student construct a quadrilateral by performing the given construction steps.

   a. Extend $\overline{BC}$ through $B$ by $b$ units and call the endpoint $D$.

   b. Draw a segment, extending to the left of D, that has length $a$ and makes a right angle with $\overline{BD}$. Call the new endpoint of this segment $E$.

   c. Draw $\overline{EA}$.

   d. One student draws $\overline{EC}$. The other student draws $\overline{EB}$.

4. Have the student who drew $\overline{EC}$ answer the following questions.

   a. The area of the quadrilateral $ACDE$ can be determined by adding together the areas of what two triangles?

   b. What is the length of $\overline{DC}$?

   c. How is the height of $\triangle AEC$ related to the height of $\triangle EDC$?

   d. Find the area of $\triangle EDC$. Simplify the expression, if necessary.

   e. Find the area of $\triangle AEC$. Simplify the expression, if necessary.

   f. Find the area of the quadrilateral $ACDE$. Simplify, if necessary.

5. Have the student who drew $\overline{EB}$ answer the following questions.

   a. The area of the quadrilateral $ACDE$ can be determined by adding together the areas of what three triangles?

   b. What is the length of $\overline{EB}$?

   c. Find the area of $\triangle ABC$.

   d. Find the area of $\triangle ABE$. NOTE: $m\angle ABE = 90°$

   e. Find the area of $\triangle EDB$.

   f. Find the area of the quadrilateral $ACDE$.

6. Since each student found the area of the same quadrilateral, the two formulas are equivalent. Have the pair equate their formulas and combine like terms.

7. Have the pair state their result.

**Assessment Goals:**

• Develop construction techniques.

• Introduce the concept of a proof.

**RESEARCH PROJECT**   *(Use after Lesson 10.4)*

As was noted in the exercises of Lesson 10.4, symmetry often occurs in nature. Research the anatomy of a frog. Encyclopedias or biology books are good sources. Locate the following parts of a frog's anatomy and determine whether or not it is symmetric.

Note: It may not be possible to determine exact symmetry, so your answers are estimations.

**a.** Aortic arches      **f.** Dorsal aorta

**b.** Kidneys      **g.** Renal portal vein

**c.** Stomach      **h.** Pharynx

**d.** Bladder      **i.** Small intestine

**e.** Large Intestine      **j.** Heart

**Assessment Goal:**

- Develop symmetry recognition.

**CONSTRUCTION**   *(Use after Lesson 10.5)*

1. Divide the class into pairs.

2. Have each pair use a ruler to draw five examples of isosceles triangles.

3. Have the pairs discuss the following.

     **a.** Describe the symmetric property of an isosceles triangle.

     **b.** Draw the lines of symmetry. Measure the base of the isosceles triangle and the bases of the right triangles formed by the line of symmetry. What is the relationship between these two lengths?

4. Given the figure, have the students answer the following questions.

     **a.** How many isosceles triangles can you draw with the base *AB*?

     **b.** Where would the third vertex (or vertices) lie?

     **c.** Given any line segment, you could draw an isosceles triangle by doing what?

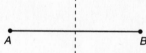

5. Have the pairs bisect a segment by performing the following steps.

     **a.** Draw $\overline{AB}$.

     **b.** Place the point of a compass on *A*. Draw an arc directly above and below $\overline{AB}$.

     **c.** Place the point of a compass on *B*. Draw an arc directly above and below $\overline{AB}$.

     **d.** With a ruler, connect the two points where the arcs intersect.

6. Give the pairs three different segments. Have them draw two isosceles triangles from each segment using the discussed construction technique.

**Assessment Goals:**

- Develop construction skills.

- Reinforce geometry terminology.

**COOPERATIVE LEARNING**   *(Use after Lesson 11.1)*

**1.** Divide the class into groups of four.

**2.** Give each group four copies of the figure below.

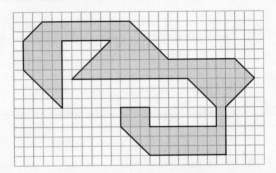

**3.** Have each student work alone to find the area of the figure. Have each student mark the individual polygons he or she used.

**4.** Have the group meet and answer the following questions.

**a.** Did everyone in the group get the same answer? If not, what was the most common answer? Examine any discrepancies.

**b.** Did everyone arrive at their answer in the same way?

**5.** Have each member explain their solution to the group.

**Assessment Goals:**

• Develop an appreciation for the idea of multiple solution methods.

• Develop calculating area skills.

**COOPERATIVE LEARNING**   *(Use after Lesson 11.7)*

**1.** Divide the class into groups of four.

**2.** Give the group the following pairs of similar figures.

Student 1:

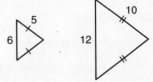

Student 2:

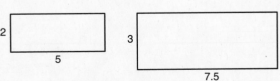

Student 3: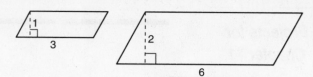

Student 4: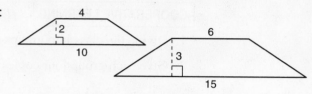

3. Have each student answer the following questions.

   a. What is the scale factor of your similar figures?

   b. Find the areas of your figures.

   c. What is the ratio of the areas of the figures?

   d. How does this ratio compare to the scale factor?

4. Have the group come together. Have each student present their result to the group.

5. Have the group discuss and answer the following questions.

   a. Hypothesize about the relationship between the scale factor of similar figures and the ratio of the figures' areas.

   b. Given two general similar triangles (one with base $b$ and height $h$ and the other with base $a$ and height $c$), show that your hypothesized relationship is still true.

   c. Given two general similar rectangles (one with length $l$ and width $w$ and the other with length $x$ and width $y$), show that your hypothesized relationship is still true.

   d. Given two general similar parallelograms (one with base $b$ and height $h$ and the other with base $a$ and height $c$), show that your hypothesized relationship is still true.

   e. Given two general similar trapezoids (one with bases $a$ and $b$ and height $h$ and the other with bases $c$ and $d$ and height $f$), show that your hypothesized relationship is still true. (HINT: If $a/c = b/d$, then $(a + b)/(c + d) = a/c = b/d$.)

**Assessment Goals:**

- Examine similar figures.

- Develop generalization skills.

## Projects for Chapter 12

**PROBLEM SOLVING** *(Use after Lesson 12.1)*

James Gregory (1638–1675) developed a method for approximating $\pi$. Gregory's approximation method made use of the sum of a sequence of numbers: $1, -\frac{1}{3}, \frac{1}{5}, -\frac{1}{7}, \frac{1}{9}, \ldots$

1. What is the pattern for the sign of each number in the sequence?

2. What is the pattern for the numerators of each number in the sequence?

3. What is the pattern for the denominators of each number in the sequence?

4. If 1 is considered term 1 in the sequence, the $n$th term of the sequence can be written as which of the following expressions? Explain your reasoning.

   a. $\dfrac{(1)^{n+1}}{n+1}$    b. $\dfrac{(1)^{n+1}}{2n-1}$    c. $\dfrac{(-1)^{n+1}}{2n-1}$    d. $\dfrac{(-1)^{n+1}}{2n+1}$

5. Gregory's approximation method for computing the decimal form of $\pi$ is $\pi \approx 4\left(1 - \frac{1}{3} + \frac{1}{5} - \frac{1}{7} + \frac{1}{9} - \ldots + n\text{th term}\right)$. Use this method to approximate $\pi$ when $n = 6$, $n = 15$, and $n = 30$.

6. Use your calculator to get an approximation of $\pi$. How close are Gregory's approximations?

**Assessment Goal:**

- Investigation of $\pi$.

**JOURNAL ENTRY** *(Use after Lesson 12.7)*

There are many solids which are made of several types of the solids or portions of the solids discussed in Chapter 12. Two examples are a capsule and a corner curio cabinet.

1. Use the terminology of the chapter to describe these solids.

2. Write a volume formula for each.

3. Find two other objects made up of several solids or portions of solids discussed in Chapter 12.

4. Write a volume formula for each.

**Assessment Goals:**

- Develop logical thinking.

- Develop students' ability to apply and extend basic information.

1. Divide the class into groups of four.

2. Give each group the following problem:

   You are responsible for the design of a water tower. The water tower must hold 10,000 ft³ of water to supply the surrounding area. You must consider the cost of the tank. The greater the surface area of the tank, the more expensive the project.

3. Consider a sphere, cylinder, and a rectangular prism with a square base for the tank. Answer the following questions. Use 3.1416 for $\pi$.

   SPHERE: a. What is the radius of a sphere that will hold the proper amount of water?

   b. The surface area of a sphere is given by $S = 4\pi r^2$. Find the surface area of the proposed tank.

   CYLINDER: a. Write a formula for the volume of a cylinder, when the volume ($V$) is 10,000 ft³.

   b. Complete the table where $r$ is the radius of the cylinder, $h$ is the height, and $S$ is the surface area.

   | $r$ | 9 | 10 | 11 | 12 | 13 | 14 |
   |---|---|---|---|---|---|---|
   | $h$ | | | | | | |
   | $S$ | | | | | | |

   c. Based on this chart, what dimensions yield the smallest surface area? What is this surface area?

   PRISM: a. Write a formula for the volume of this prism, when the volume ($V$) is 10,000 ft³.

   b. Complete the table where $x$ is the length of the square base, $h$ is the height of the prism, and $S$ is the surface area.

   | $x$ | 19 | 20 | 21 | 22 | 23 | 24 |
   |---|---|---|---|---|---|---|
   | $h$ | | | | | | |
   | $S$ | | | | | | |

   c. Based on this chart, what dimensions yield the smallest surface area? What is this surface area?

4. What is the most cost effective shape of the water tank?

**Assessment Goals:**

• Develop problem solving skills.

• Develop skills in using volume and surface area formulas.

**GRAPHING CALCULATOR**  (*Use after Lesson 13.5*)

The following instructions apply to a TI-82 by Texas Instruments. The procedure may be adapted to other graphing calculators with statistical capabilities.

1. Given the table at the right, find the equation of a line passing through these points.

| $x$ | 0 | 2 | 4 | 6 | 8 | 10 |
|---|---|---|---|---|---|---|
| $y$ | 3 | 4 | 5 | 6 | 7 | 8 |

   a. Set the window to: Xmin = −10, Xmax = 10, Xscl = 1, Ymin = −10, Ymax = 10, Yscl = 1 (See page 627 of *Passport to Algebra and Geometry*.)

   b. Press ⎡STAT⎤ ⎡ENTER⎤ (1:Edit).

   c. Under $L_1$, type in the $x$-coordinates.

   d. Use the arrow keys to move to $L_2$ and enter the $y$-coordinates.

   e. Press ⎡STAT⎤ and move cursor to CALC.

   f. Press ⎡5⎤ (LinReg (ax + b)) ⎡ENTER⎤ .

   g. $a$ is the slope and $b$ is the $y$-intercept. What is the equation of the line?

2. Use the following procedure to graph your equation and the points in the table. Then you can see whether your graph contains the given points.

   a. Press ⎡STAT PLOT⎤ (i.e. ⎡2nd⎤ ⎡Y=⎤ ).

   b. Press ⎡ENTER⎤ (1:Plot1).

   c. Enter the following: ON, Type: ⎡⠂⠂⎤ , Xlist: $L_1$, Ylist: $L_2$, and Mark ⎡□⎤ .

   d. Press ⎡Y=⎤ and enter the equation of the line.

   e. Press ⎡GRAPH⎤ .

3. To delete your lists, follow the procedure.

   a. Press ⎡STAT⎤ ⎡4⎤ (ClrLst).

   b. Enter $L_1$, $L_2$ (i.e. ⎡2nd⎤ ⎡1⎤ ⎡,⎤ ⎡2nd⎤ ⎡2⎤ ) ⎡ENTER⎤.

4. Use your calculator to approximate the equation of a line passing through the given points. Then graph the data points and equation with your calculator. Note: you may wish to adjust the range.

| $x$ | −3 | −2 | −1 | 0 | 1 | 2 | 3 |
|---|---|---|---|---|---|---|---|
| $y$ | −2.25 | −2 | −1.65 | −1.5 | −1.1 | −0.32 | −0.61 |

**Assessment Goal:**

• Develop graphing calculator skills.

**GRAPHING CALCULATOR**   *(Use after Lesson 13.5)*

**1.** Divide the class into pairs.

**2.** Give each pair a graphing calculator and have them set the range to Xmin = −5, Xmax = 5, Xscl = 1, Ymin = −5, Ymax = 5, Yscl = 1.

**3.** Have the pairs graph the two equations $y = 3x + 2$ and $y = x − 2$ simultaneously. Then have them answer the questions:

   **a.** What point looks like it is on both graphs?

   **b.** Verify that this point is a solution to both equations.

   **c.** Are there any other points that will be on both lines? Explain your answer.

**4.** Have the pairs graph $y = 2x + 1$ and $y = 2x − 3$ and answer the questions:

   **a.** Do the graphs appear parallel?

   **b.** Zoom out. Do the graphs still appear parallel?

   **c.** How can you be sure that the lines are indeed parallel?

**5.** Have the pair graph $y = −\frac{6}{7}x − \frac{5}{7}$ and $y = −\frac{4}{5}x − 2$ and answer the questions:

   **a.** Do the graphs appear parallel?

   **b.** Zoom out. Do the graphs still appear parallel?

   **c.** How can you be sure that the lines are not parallel?

**Assessment Goals:**

- Develop graphing calculator skills.

- Evaluate ability to interpret linear equations.

**COOPERATIVE LEARNING**   *(Use after Lesson 13.8)*

**1.** Divide the class into groups of four.

**2.** Give each group a state road map and a red pen (or pencil).

**3.** Have the groups draw (in red) a Cartesian Plane on the road map. The origin should be located approximately at the center of the state. The length of the units should correspond to information given in the map's key. If you prefer, the Cartesian Plane can be drawn on tracing paper and taped securely to the map.

**4.** Have the groups locate the capital city and at least four other cities. Have them identify the cities with an ordered pair.

**5.** Each member of the group then takes a turn calculating the distance from the capital to one of the other cities using the distance formula.

**6.** Have the group verify each answer by using the map's key.

**Assessment Goals:**

- Develop map reading skills.

- Evaluate students' knowledge of the distance formula.

---

**JOURNAL ENTRY** *(Use after Lesson 14.1)*

Examine the following sets of data:

Set 1: 45, 26, 32, 43, 39, 41, 38, 42

Set 2: 10, 12, 9, 8, 14, 15, 7, 6

Set 3: 3, 72, 79, 85, 74, 86, 82, 90

Set 4: 7, 12, 13, 18, 106, 15, 17, 11

For which set(s) of data is the median a better measure of central tendency than the mean? Write a paragraph stating under what circumstances the median is a better form of measure for central tendency.

**Assessment Goals:**

- Develop writing skills.

- Develop students' ability to interpret and express data.

**COOPERATIVE LEARNING** *(Use after Lesson 14.4)*

1. Divide the class into groups of three.

2. Give each group the following information.

    DEFINITION:

    Given the $2 \times 2$ matrix $\begin{bmatrix} a & b \\ c & d \end{bmatrix}$, its determinant $\begin{vmatrix} a & b \\ c & d \end{vmatrix}$ is defined to be $ad - bc$.

    CRAMER'S RULE:

    Given two linear equations of the form $Ax + By = C$ and $Dx + Ey = F$, you can find the point that lies on both lines by using determinants.

    $$x = \frac{\begin{vmatrix} C & B \\ F & E \end{vmatrix}}{\begin{vmatrix} A & B \\ D & E \end{vmatrix}} \qquad y = \frac{\begin{vmatrix} A & C \\ D & F \end{vmatrix}}{\begin{vmatrix} A & B \\ D & E \end{vmatrix}}$$

3. Use Cramer's Rule to find a solution to both equations. Verify your solution by substituting into both original equations.

    a. $2x - y = -1$ and $3x - y = 2$

    b. $x - 5y = 4$ and $-2x + y = 1$

    c. $3x - 2y = 4$ and $x + 4y = 3$

4. Have the group discuss what happens to Cramer's Rule if the two given lines are parallel.

**Assessment Goal:**

- Expose students to an application of matrices.

## GRAPHING CALCULATOR  *(Use after Lesson 14.5)*

1. Divide the class into pairs.

2. Provide each pair with a graphing calculator.

3. Have the students examine the graphs of $y = 2$, $y = -3$, and $y = \frac{7}{2}$ and answer the questions.

   **a.** Make up three more polynomial equations of the form $y = $ a number.

   **b.** What do all of these graphs have in common?

4. Examine the graphs of $y = 2x + 1$, $y = -4x + 2$, $y = \frac{1}{2}x - 2$, and of similar equations. What do all the graphs have in common?

5. Examine the graphs of $y = x^2 + 1$, $y = 3x^2 - 2$, $y = -2x^2$, and of similar equations. What do all the graphs have in common?

### Assessment Goals:

- Develop graphing calculator skills.
- Familiarize students with graphs.

## COOPERATIVE LEARNING  *(Use after Lesson 14.8)*

1. Divide the class into groups of four.

2. Have the groups discuss the following example which extends the multiplying technique described in Lesson 14.8 to polynomials with more than two terms.

$$(x + 1)(x^2 + 2x + 1) = (x + 1)[(x^2 + 2x) + 1]$$
$$= (x + 1)(x^2 + 2x) + (x + 1) \cdot 1$$
$$= x^3 + 2x^2 + x^2 + 2x + x + 1$$
$$= x^3 + 3x^2 + 3x + 1$$

3. Have the group perform the multiplications in b–d.

   **a.** $(x + 1)^1$

   **b.** $(x + 1)^2 = (x + 1)(x + 1)$

   **c.** $(x + 1)^3 = (x + 1)(x + 1)(x + 1)$

   **d.** $(x + 1)^4 = (x + 1)(x + 1)(x + 1)(x + 1)$

4. Complete the table.

| Step | Coefficients of the terms in the product |
|------|------------------------------------------|
| $(x + 1)^1$ | 1, 1 |
| $(x + 1)^2$ | 1, 2, 1 |
| $(x + 1)^3$ | |
| $(x + 1)^4$ | |

5. Do these numbers follow a pattern you have seen before? If so, what? (HINT: Review Lesson 8.7)

6. Without doing the multiplication, find $(x + 1)^8$.

### Assessment Goals:

- Develop polynomial multiplication skills.
- Develop students' ability to recognize patterns.

*Passport to Algebra and Geometry*

**WHAT ARE PARTNER QUIZZES?**

Partner quizzes are used not only to assess students' achievement, but also to encourage students to communicate mathematically with each other. Copymasters for Mid-Chapter Partner Quizzes are provided on pages 44–57. They correspond in content to the Mid-Chapter Assessments in the student's text and to the Mid-Chapter Tests in the *Formal Assessment* copymasters. Here are two ways to use partner quizzes.

**FIRST WAY**

Three copies of the *same* quiz are given to two students. For the first ten minutes, each person works independently, answering as many of the questions as possible. The two students are then allowed to work together as partners. During this time, they are to complete the third copy of the quiz. The two partners turn in only the third copy, which shows the answers and work that they have agreed upon.

This type of quiz emphasizes the importance of mathematical communication. A student whose answers differ from his or her partner's needs to be able to explain how he or she approached the problem and also to explain the reasoning used to make each decision. Students need to recognize that there is often more than one correct approach to solving a problem. The partners then need to decide which approach they will submit on the third copy of the quiz. A student who develops a mental block on a problem can benefit from having his or her memory jogged by a partner.

**SECOND WAY**

Another way to give students a partner quiz is to have one problem written on the overhead. One student (the interpreter) in each pair faces the overhead and the other (the solver) faces away from the overhead. Ask each interpreter to silently read and interpret the problem. After a couple of minutes, turn off the overhead. Each interpreter then explains the problem to his or her solver. The solver must then use the information given by the interpreter to solve the problem. After the interpreter has explained the problem, he or she must remain quiet. The interpreter cannot help his or her partner solve the problem. Therefore, it is important for the interpreter to do a good job explaining the problem and giving all of the relevant information. The partners may then switch roles and continue with the next problem or they may wait until the next quiz to switch roles.

This type of activity emphasizes the importance of reading the problem, and then interpreting the information. It also helps to teach students to differentiate between information needed to solve the problem and extra information.

1. Find the next three numbers in the sequence. (1.1)

   2, 5, 11, 23, ☐?☐, ☐?☐, ☐?☐

   1. _____

2. Draw the next figure. (1.1)

   2. _____

3. Name three different symbols that indicate the division operation. (1.2)

   3. _____

4. Write a symbolic description of the statement:

   The difference of 12 and 7 is 5. (1.2)

   4. _____

5. Find the quotient. $\dfrac{5142}{24}$ (1.2)

   5. _____

6. If the side of a cube has length 4 inches, find the volume of the cube. (1.3)

   6. _____

7. A square piece of carpet has an area of 231.04 ft². What is the perimeter of a room in which the carpet fits wall to wall? (1.3)

   7. _____

8. Evaluate the expression without using a calculator. (1.4)

   $13 - (2 + 1)^2 + 4 - 2$

   8. _____

9. Insert parentheses to make the statement true. (1.4)

   $21 \div 3 + 4 \times 2 = 22$

   9. _____

10. You go to the mall to shop for items to take on your vacation to the beach. You purchase 3 paperback books for $4.95 each, 2 cassette tapes for $12.95 each, a bottle of sunscreen for $2.75, and sunglasses for $10. How much money did you spend? (1.4)

    10. _____

1. Use the Distributive Property to rewrite the expression $2(x + y + 5)$.
   (2.1)

   1. _____

2. You are a running back on a football team. During the first 3 games
   you scored 1, 2, and 1 touchdowns (6 points each), respectively.
   How many points did you score during these three games?   (2.1)

   2. _____

3. Simplify the expression $6a + 2(a + 3) + 1 + a5$.   (2.2)

   3. _____

4. Write an expression for the area
   of the figure. What is the area
   if $x = 2$ inches?   (2.2)

   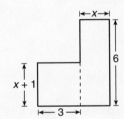

   4. _____

5. Which of the following numbers is a solution to the equation
   $x^2 - 2x + 1 = 4$?   (2.3)

   a. $x = 2$          b. $x = 3$          c. $x = 4$

   5. _____

6. Solve the equation using mental math.   (2.3)

   $12 - x = 8$

   6. _____

7. Your entire family went to the movies. There are 5 people in your
   family and the total cost was $30. How much did each ticket cost?
   (2.3)

   7. _____

8. Solve the equation $2.431 = x - 4.722$.   (2.4)

   8. _____

9. In 6 years Maria will be able to vote. How old is Maria now?   (2.4)

   9. _____

10. The bill at a restaurant is $25.32. You leave $29.15. How much did you
    leave for a tip?   (2.4)

    10. _____

1. Draw a number line and plot the integers $-3, 2, 0,$ and $-1$.   (3.1)

   1. _____

2. Compare the integers using $<$, $>$, or $=$.   (3.1)

   $|-12|$ $\boxed{?}$ $6$

   2. _____

3. What is the sum of any two opposites?   (3.2)

   3. _____

4. Find the sum.   (3.2)

   $-10 + |-3|$

   4. _____

5. Simplify the expression. Then evaluate the expression when $x = 4$.
   (3.3)

   $-3x - 12 + 8x + 4$

   5. _____

6. Write an equation that models the given directions: Begin at 3,
   move 4 units left, move 2 units right, and move 1 unit left. Evaluate.
   (3.3)

   6. _____

7. Subtracting $a$ from $b$ is the same as adding what two numbers?
   (3.4)

   7. _____

8. Complete the statement with *always, sometimes,* or *never.*   (3.4)

   $a - (-b)$ is $\boxed{?}$ positive.

   8. _____

9. Evaluate $-a^2$ when $a = -3$.   (3.5)

   9. _____

10. Complete the statement with *always, sometimes,* or *never.*   (3.5)

    The product of two numbers with the same signs is $\boxed{?}$ negative.

    10. _____

1. What is the inverse operation of division?   (4.1)

1. _____

2. Solve the equation. Check your solution.   (4.1)

   $-2s + 7 = 35$

2. _____

3. A medium pizza with two toppings costs $9 plus $2 for each additional topping. You have $13. How many extra toppings can you get? Write an algebraic model and answer the question.   (4.1)

3. _____

4. Solve the equation. Check your solution.   (4.2)

   $3x - 2x + 3 - 5x = 11$

4. _____

5. The sum of the measures of two angles is 90°. The first angle's measure is 6° more than twice the measure of the second angle. Find the measure of the second angle.   (4.2)

5. _____

6. What is the reciprocal of $-\frac{2}{3}$?   (4.3)

6. _____

7. The temperature, $T$, in Celsius, of a solution is modeled by $T = 0.4t + 86$ where $t$ is time in minutes and $t = 0$ corresponds to 1:00 P.M. When does the solution boil (100°C)?   (4.3)

7. _____

8. Write an equation that represents the verbal sentence. Solve the equation.   (4.4)

   4 times the sum of a number and 2 is 12.

8. _____

9. Solve the equation $18 = 2(x - 3) + 26$.   (4.4)

9. _____

10. Write an equation for the perimeter of the given triangle. Solve for $x$.   (4.4)

10. _____

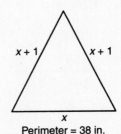

$x + 1$   $x + 1$

$x$

Perimeter = 38 in.

In Exercises 1–3, use the picture graph at the right. The graph shows the populations of the five counties of Rhode Island.   (5.1)
*(Source: The World Almanac and Book of Facts 1996)*

Counties of Rhode Island        Population

Bristol            = 50,000 people

Kent

Newport

Providence

Washington

1. Estimate the population of Providence county.

1. _____

2. Estimate the population of Rhode Island.

2. _____

3. Which has the larger population, Providence or the rest of the state?

3. _____

In Exercises 4–6, use the table at the right. The table lists the life expectancy of males and females born from 1920 through 1990.   (5.3) *(Source: The World Almanac and Book of Facts 1996)*

| Year | Male | Female | Year | Male | Female |
|------|------|--------|------|------|--------|
| 1920 | 53.6 | 54.6   | 1960 | 66.6 | 73.1   |
| 1930 | 58.1 | 61.6   | 1970 | 67.1 | 74.7   |
| 1940 | 60.8 | 65.2   | 1980 | 70.0 | 77.5   |
| 1950 | 65.6 | 71.1   | 1990 | 71.8 | 78.8   |

4. Create a double line graph for the data.

4. _____

5. What is the trend in life expectancy for both males and females?

5. _____

6. During which year was the life expectancy of males closest to that of females?

6. _____

7. Choose a type of graph that best represents the following data. Draw the graph. The table shows the number of hospitals in the four most populous cities of New York.   (5.2, 5.4) *(Source: The World Almanac and Book of Facts 1996)*

7. _____

| City | Hospitals |
|------|-----------|
| Buffalo | 20 |
| New York City | 81 |
| Rochester | 8 |
| Yonkers | 3 |

1. Find all the factors of 132.   (6.1)

1. _____

2. The volume of a box is 60 in.². The length, width, and height of the box are natural numbers. Find all of the possible dimensions of the box.   (6.1)

2. _____

3. Write the prime factorization of 1260. Write your answer in exponent form.   (6.2)

3. _____

4. Complete the statement with *always, sometimes,* or *never.* A prime number is ⬚?⬚ even.   (6.2)

4. _____

5. Find the greatest common factor of $24x^2y^3$, $4x^2y^4$, $8x^5y^2$, and $12x^3y^6$.   (6.3)

5. _____

6. Find a number which is relatively prime with 30.   (6.3)

6. _____

7. Find the least common multiple of $8xy$, $3x^2$, and $4xy^3$.   (6.4)

7. _____

8. Write three fractions that are equivalent to $\frac{12}{15}$.   (6.5)

8. _____

9. Which fraction is not equivalent to the rest of the group?   (6.5)

$\frac{63}{54}, \frac{14}{12}, \frac{56}{46}, \frac{35}{30}, \frac{42}{36}$

9. _____

10. Write the fractions from greatest to least.   (6.5)

$\frac{2}{5}, \frac{4}{9}, \frac{5}{11}, \frac{9}{22}$

10. _____

Names _____

_____

1. Write an expression for the geometric model and evaluate. (7.1)

 +  –

1. _____

2. Add and simplify, if possible. (7.1)

$$\frac{-3a}{2b} + \frac{7a}{2b}$$

2. _____

3. Find the sum. Then simplify. (7.2)

$$\frac{1}{3} + \frac{5}{12} + \frac{3}{8}$$

3. _____

4. Solve the equation. (7.2)

$$\frac{2}{3} + c = \frac{1}{5}$$

4. _____

5. Complete the following statement with *always, sometimes,* or *never.* The least common denominator of two fractions is ☐? the product of the two denominators. (7.2)

5. _____

6. Evaluate the expression by first rewriting in decimal form. Round the result to three decimal places. (7.3)

$$\frac{3}{7} + \frac{1}{6} - \frac{2}{9}$$

6. _____

7. The circle graph compares different uses of your allowance money. What portion of your allowance goes toward your college fund? Round your results to two decimal places. (7.3)

7. _____

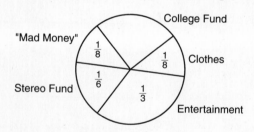

8. Multiply and simplify, if possible. (7.4)

$$-\frac{3}{8} \cdot \frac{4}{5}$$

8. _____

9. How does multiplying a negative number by a fraction less than 1 affect the size of the original number? (7.4)

9. _____

10. Find the area of the given figure. (7.4)

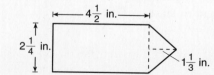

10. _____

1. Is $\dfrac{12 \text{ households}}{19 \text{ households}}$ a ratio or a rate?   (8.1)

1. _____

2. A 6.4 ounce tube of toothpaste costs $1.83. A 5.2 ounce tube of toothpaste costs $1.54. Which is the better buy? (8.1)

2. _____

3. On a committee of 10 people, a resolution passed unanimously. Write this verbal phrase as a ratio.   (8.1)

3. _____

4. Solve the proportion for $a$.   $\dfrac{4}{a} = \dfrac{6}{5}$   (8.2)

4. _____

5. In the given figure, $\triangle ABC$ is similar to $\triangle DEF$. Find side $b$.   (8.2)

5. _____

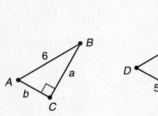
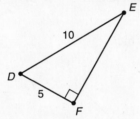

6. A model of a flying saucer used in a science fiction movie is used to depict a life-sized saucer. The model is made on a 1-to-125 scale. The saucer in the film appears to have a 210-foot diameter. What is the diameter of the model?   (8.3)

6. _____

7. If a secretary can type an average of 83 words per minute, how many words can the secretary type in 5 minutes?   (8.3)

7. _____

8. What is 12% of 26?   (8.4)

8. _____

9. After spending $43.10 on a meal, you leave a $6.00 tip. Approximately what percent tip did you leave?   (8.4)

9. _____

10. The Pittsburgh Pirates have won 56% of their games so far this season. They have won 42 games. How many games have they played?   (8.4)

10. _____

1. Write an algebraic equation for the sentence and approximate the solutions to three decimal places. The difference of $x$ squared and 4 is 8.   (9.1)

1. _____

2. How long are the legs of the given right triangle if the area is 98 in²?   (9.1)

2. _____

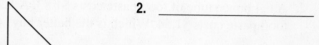

3. Which number is larger, $\sqrt{\frac{5}{8}}$ or $\frac{4}{5}$?   (9.2)

3. _____

4. Is $2.03\overline{2}$ a rational or irrational number?   (9.2)

4. _____

5. The hypotenuse of a right triangle has length 26. One leg has length 10. Find the length of the other leg.   (9.3)

5. _____

6. The sides of a right triangle have lengths 36, 39, and 15. Which length corresponds to the hypotenuse?   (9.3)

6. _____

7. Write the Pythagorean Theorem for the given triangle.   (9.3)

7. _____

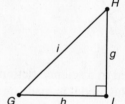

**Use the figure to answer Exercises 8–10.   (9.4)**

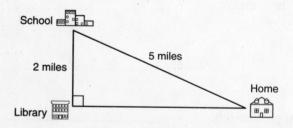

8. What is the distance from your house to the library? Round your solution to 1 decimal place.

8. _____

9. Every day you ride from home, to school, to the library, and then home. What is your round trip distance?

9. _____

10. What is the area of this section of your town?

10. _____

**In Exercises 1 and 2, use the diagram below.** (10.1)

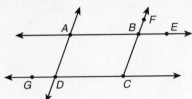

1. At what point does $\overleftrightarrow{AE}$ intersect with $\overrightarrow{CF}$?

1. _____

2. What two lines appear to be parallel?

2. _____

**In Exercises 3 and 4, use the diagram at the right.** (10.2)

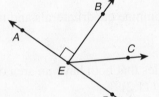

3. List all acute angles.

3. _____

4. List all obtuse angles.

4. _____

**In Exercises 5 and 6, use the diagram at the right.** (10.3)

5. If $m\angle 1 = 32°$, find $m\angle 12$.

5. _____

6. List all angles congruent to $\angle 1$.

6. _____

7. Complete the figure so that it has a horizontal line of symmetry. (10.4)

7.

8. Draw a figure that will coincide with itself after being rotated 45°, 90°, 135°, or 180° in either direction. (10.4)

8. _____

9. If the obtuse angle of an isosceles obtuse triangle has measure 142°, what are the measures of the other angles? (10.5)

9. _____

10. If $\triangle ABC$ is an equilateral triangle, find the measures of $\angle A$, $\angle B$, and $\angle C$. (10.5)

10. _____

**In Exercises 1 and 2, use the figure below.** **(11.1)**

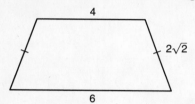

4

$2\sqrt{2}$

6

1. Find the area of the given isosceles trapezoid.

1. _____

2. Find the perimeter of the given isosceles trapezoid.

2. _____

3. Find three different ways to combine quadrilaterals and triangles to form an octagon. (11.1)

3. _____

4. Find an example of two triangles that both have an area of 12 square units, but are not congruent. (11.2)

4. _____

5. The two triangles below are congruent. Label the vertices of the second triangle *X, Y,* and *Z* so that $\triangle ABC \cong \triangle XYZ$. (11.2)

5.

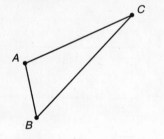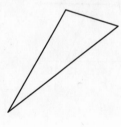

C

A

B

6. The vertices of $\triangle ABC$ are $A(0, 2)$, $B(3, 1)$, and $C(4, 4)$. Find the vertices of $\triangle A'B'C'$ when $\triangle ABC$ is reflected across the *y*-axis. (11.3)

6. _____

7. Reflect the Roman numerals (I - X) about a vertical line positioned to the right of the numeral. Which numerals remain the same? Which turn into new numerals? (11.3)

7. _____

8. The vertices of $\triangle ABC$ are $A(1, 0)$, $B(4, 0)$, $C(2, 2)$. Find the vertices of $\triangle A'B'C'$ when $\triangle ABC$ is rotated 180° about the point (0, 0). (11.4)

8. _____

9. A quadrilateral has vertices $A(1, 1)$, $B(3, 2)$, $C(5, 1)$, and $D(2, -2)$. Find the vertices of the image after it has been translated one unit down and two units left. (11.5)

9. _____

10. Write a motion rule for translating a figure four units up and three units left. (11.5)

10. _____

**In Exercises 1 and 2, use the figure at the right.** (12.1)

Area of shaded region is $6\pi$ in.$^2$

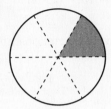

1. Find the radius of the circle.

2. Find the circumference of the circle.

3. Find the diameter of a circle with circumference 7 inches. (12.1)

4. Find the perimeter of the figure. Use 3.14 for $\pi$. (12.1)

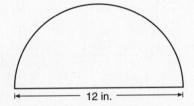

5. If a pyramid has 6 faces, how many of these faces share a common point? What is the base of the pyramid? (12.2)

6. Which two solids have a circle as a base? (12.2)

7. Which of the cylinders has the greatest surface area? (12.3)

a.

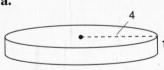

b.

8. What is the surface area of the prism formed by the given net? (12.3)

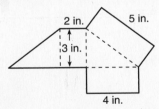

9. Write a formula for the volume of a cube in terms of $s$, the length of the cube's side. (12.4)

10. Solve for $x$.   $V = 70$ in.$^3$ (12.4)

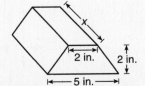

1. _____

2. _____

3. _____

4. _____

5. _____

6. _____

7. _____

8. _____

9. _____

10. _____

**1.** List three solutions of $3x + 4y = 10$.   (13.1)

1. _____

**2.** Write a linear equation that shows the relationship between $x$ and $y$ in the figure.   (13.1)

2. _____

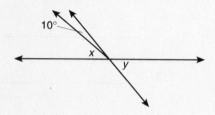

**3.** What do all of the solutions of the equation $x = 4$ have in common?   (13.2)

3. _____

**4.** Sketch the graph of the equation $y = \frac{1}{3}x + 1$.   (13.2)

4. _____

**5.** Find the $x$-intercept of the line given by $y = \frac{2}{3}x - 5$.   (13.3)

5. _____

**6.** What is the only point that can be both an $x$-intercept and a $y$-intercept?   (13.3)

6. _____

**7.** In an experiment, you record the temperature of a solution at one minute intervals. The graph is below. What does the $t$-intercept represent?   (13.3)

7. _____

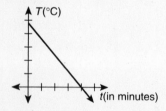

**8.** Find the slope of the line passing through the points $(-3, 2)$ and $(5, 3)$.   (13.4)

8. _____

**9.** Line 1 passes through $(2, 6)$ and $(4, 4)$. Line 2 passes through $(-4, 8)$ and $(-6, 6)$. Are the lines parallel?   (13.4)

9. _____

**10.** Estimate the slope of the graph.   (13.4)

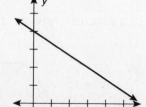

10. _____

**In Exercises 1–5, use the following test score data.**

80, 69, 51, 97, 85, 77, 5, 95, 92, 72, 71, 89, 64, 81, 85

1. Use a stem-and-leaf plot to order the data.   (14.2)

1. _____

2. Find the mean.   (14.1)

2. _____

3. Find the median.   (14.1)

3. _____

4. Find the mode.   (14.1)

4. _____

5. Which measure of central tendency test describes the data?   (14.1)

5. _____

6. The following box-and-whisker plot shows the test scores for a 100-point test. If there are 10 students in the class, write two sets of test scores that could be represented by the box-and-whisker plot.   (14.3)

6. _____

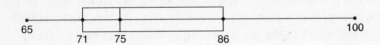

7. Write the table as a matrix.   (14.4)

7. _____

|  | High Temp. | Low Temp. |
|---|---|---|
| Monday | 93° | 72° |
| Tuesday | 90° | 70° |
| Wednesday | 88° | 75° |
| Thursday | 94° | 74° |
| Friday | 89° | 69° |
| Saturday | 85° | 66° |
| Sunday | 87° | 65° |

8. Evaluate.   (14.4)

8. _____

$$\begin{bmatrix} 2 & 3 & 1 \\ -1 & 4 & 0 \\ 3 & 2 & -2 \end{bmatrix} + \begin{bmatrix} 1 & 0 & -2 \\ 3 & -5 & 1 \\ 0 & 0 & 6 \end{bmatrix} - \begin{bmatrix} -4 & 2 & 1 \\ 1 & 2 & -3 \\ 5 & 6 & -1 \end{bmatrix}$$

**WHAT IS GROUP ASSESSMENT?**

Students learn from each other when they work together. In a small group setting, students are better able to share their ideas while exploring a problem, developing a strategy to solve the problem, and then solving the problem. Assessment should be part of this learning activity; it provides insight into the thought process of the students.

The goal of the group assessment in this supplement is not to grade the students on the solution. The goal is to assess (1) their understanding of the problem, (2) how they solve the problem, and (3) how well they work together in a small group.

The blackline masters for group assessment can be found on pages 59–86 of this supplement. There are two pages for each chapter. The first page is an activity or a problem for a group of three or four students to explore. The second page is a similar activity or problem for the individual to explore. The individual page should motivate students to participate and work to understand the group activity.

**GRADING**

There are two ways to assign grades to the students. The group and individual pages are scored and each member of the group receives the sum of the points on the group page and their individual page. Alternatively, the group and individual pages are scored and each member of the group receives the sum of the points on the group page and the average of the individual pages of the members of the group. When using the second grading option, students will often make a greater effort in the group setting to insure that every person understands the problem and the solution.

**Materials:** None

**Explore:** The von Koch snowflake is constructed by taking a line segment and putting in a triangular "notch". A notch is then put in each segment of the new figure. Step 0 of the von Koch snowflake has 1 segment: ——————— . Step 1 has 4 segments: _____/_____ . Find an algebraic model for the number of segments in each step.

**1.** Develop a plan to solve the problem.

**2.** How many segments are there in step 2 and step 3?

**3.** Use your plan to determine the number of segments in the *n*th step.

**4.** What is the advantage of having an algebraic model to find the number of segments for any given step?

**Materials:**   None

**Explore:**   A variation of the von Koch snowflake is constructed by taking
a line segment and inserting a square "notch". A notch is then
inserted in each segment of the new figure. Step 0 has 1 segment:
_____ . Step 1 has 5 segments: _____⌐‾⌐_____ .
Find an algebraic model for the number of segments in the
$n$th step.

**1.** Develop a plan to solve the problem.

**2.** How many segments are there in step 2 and step 3?

**3.** Use your plan to determine the number of segments in the $n$th step.

**4.** How does your method compare to the one used in the group activity?

**Materials:** None

**Explore:** During one grading period, you will be assigned 5 essays worth 20 points apiece, 5 homework assignments worth 20 points apiece, and 2 tests worth 50 points apiece. The grading scale for the grading period is as follows:

A: 90% – 100%
B: 80% – 89.9%
C: 70% – 79.9%
D: 60% – 69.9%
F: 0% – 59.9%

You have already turned in all of the work, except for the second test. You have earned 220 points. Find a range of test scores for which you will earn a B for the grading period.

**1.** Rewrite the problem in your own words.

**2.** Develop a plan for solving the problem.

**3.** Write an algebraic model(s).

**4.** What is the range of test scores that will lead to a B grade?

**Materials:** None

**Explore:** During one grading period, you will be assigned 6 essays worth 25 points apiece, 5 homework assignments worth 20 points apiece, and 2 tests worth 75 points apiece. The grading scale for the grading period is as follows:

A: 90% – 100%
B: 80% – 89.9%
C: 70% – 79.9%
D: 60% – 69.9%
F: 0% – 59.9%

You have already turned in all of the work, except for the second test. You have earned 290 points. Find a range of test scores for which you will earn a B for the grading period.

1. Rewrite the problem in your own words.

2. Develop a plan for solving the problem.

3. Write an algebraic model(s).

4. What is the range of test scores that will lead to a B grade?

5. Is it possible for you to earn an A for the grading period? If so, how?

6. Did you use the same plan as the group? If not, explain why you used your method.

**Materials:**    Graph paper

**Explore:**    Examine the graphical representation of $y = |x|$.

1. Without constructing a table, in what quadrants do you think all of the points will lie?

2. Construct a table that lists several solutions to the equation. Plot the points.

3. Describe the pattern made by the solution points.

4. What do you notice about the $y$-coordinates associated with $x$-coordinates that differ only in their sign? Using terminology from Lesson 3.1, explain why this occurs.

**Materials:**   Graph paper

**Explore:**   Examine the graphical representation of $y = |x - 1|$.

1. Without constructing a table, in what quadrants do you think all of the points will lie?

2. Construct a table that lists several solutions to the equation. Plot the points.

3. Describe the pattern made by the solution points. How does it differ from the pattern in the group activity?

4. Do opposite values of $x$ produce the same $y$ value? If yes, explain why this occurs. If no, is there some other pattern that leads to equal $y$ values?

**Materials:**   None

**Explore:**   An isosceles trapezoid is a quadrilateral in which exactly one
pair of opposite sides are parallel and the other pair of opposite
sides have equal length. Find a formula for the area of an
isosceles trapezoid.

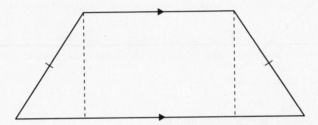

1. Label the parallel sides of the given isosceles trapezoid *a* (shorter side)
   and *b* (longer side). Label the height of one of the indicated triangles as
   *h*. (This is also known as the height of the trapezoid.) Label the length
   of the base of one of the indicated triangles as *c*.

2. Find a formula for the area of one of the triangles in terms of *c* and *h*.
   Will the formula be the same for the other triangle? Explain your
   answer.

3. Find a formula for the area of the indicated rectangle.

4. Write a verbal and algebraic model for the area of the isosceles
   trapezoid.

5. Explain why the base of one of the indicated triangles can be written as
   $\frac{1}{2}(b - a)$.

6. Substitute $\frac{1}{2}(b - a)$ for *c* in your algebraic model of the area of an
   isosceles trapezoid and simplify your answer.

7. Another way to find the formula for the area of an isosceles trapezoid
   involves cutting off one of the triangles and rearranging the pieces so
   that it is a rectangle. What is the length and height of the rectangle?
   What is the area? Which method for finding the area of an isosceles
   trapezoid do you prefer? Why?

**Materials:** None

**Explore:** A trapezoid is a quadrilateral in which exactly one pair of its opposite sides are parallel. Find a formula for the area of a trapezoid.

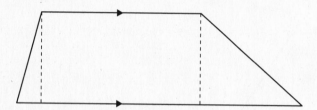

1. Label the parallel sides of the given trapezoid *a* (shorter side) and *b* (longer side). Label the height of one of the indicated triangles as *h*. (This is also known as the height of the trapezoid.) Label the base of the indicated triangle on the left as *c*. Label the base of the indicated triangle on the right as *d*.

2. Find the formulas for each of the indicated triangles. Will the formulas be the same? Explain your answer.

3. Find a formula for the indicated rectangle.

4. Write a verbal and an algebraic model for the area of the trapezoid.

5. Explain why $a + c + d = b$.

6. Solve $a + c + d = b$ for *c* and substitute it into your formula for the area of a trapezoid. Simplify the formula.

7. Compare your result with the formula for the area of an *isosceles* trapezoid found in the group exercise.

8. Would the cutting and rearranging method used in the group assessment work in this general case? Explain your answer.

**Materials:** Graph paper, rulers

**Explore:** You are the campaign manager for the mayor in your city who is running for re-election. The following data about your city will be displayed during a debate with the opponent. What types of graphs will best represent the data?

**Number of Crimes per 100,000 People**

| Year | No. of Crimes | Year | No. of Crimes |
|------|------|------|------|
| 1988 | 6200.4 | 1991 | 6103.5 |
| 1989 | 6175.3 | 1992 | 6100.2 |
| 1990 | 6140.7 | 1993 | 6084.7 |

(Note: Number of crimes in your state per 100,000 people in 1993 was 5660.)

**Average Number of Unemployed per 100 People**

| Year | No. of Unemployed | Year | No. of Unemployed |
|------|------|------|------|
| 1988 | 6.2 | 1991 | 5.4 |
| 1989 | 6.0 | 1992 | 5.2 |
| 1990 | 5.7 | 1993 | 5.1 |

(Note: An average of 7.2 per 100 people were unemployed in your state in 1993.)

1. Which set of data should the mayor emphasize? Why?

2. Discuss what type of graph would be the best way to present this positive information.

3. Draw the graph.

4. Discuss strategies for making the other set of data appear positive.

5. What type of graph will you use to represent this negative data? Why?

6. Draw the graph.

**Materials:** Graph paper, rulers

**Explore:** You are the campaign manager for the opponent in your city's upcoming mayoral election. The following data about your city will be displayed during a debate with the mayor. What types of graphs will best represent the data?

**Number of Crimes per 100,000 People**

| Year | No. of Crimes | Year | No. of Crimes |
|------|------|------|------|
| 1988 | 6200.4 | 1991 | 6103.5 |
| 1989 | 6175.3 | 1992 | 6100.2 |
| 1990 | 6140.7 | 1993 | 6084.7 |

(Note: Number of crimes in your state per 100,000 people in 1993 was 5660.)

**Average Number of Unemployed per 100 People**

| Year | No. of Unemployed | Year | No. of Unemployed |
|------|------|------|------|
| 1988 | 6.2 | 1991 | 5.4 |
| 1989 | 6.0 | 1992 | 5.2 |
| 1990 | 5.7 | 1993 | 5.1 |

(Note: An average of 7.2 per 100 people were unemployed in your state in 1993.)

1. Which set of data should you emphasize? Why?

2. Discuss what type of graph would be the best way to present this set of data.

3. Draw the graph.

4. Discuss strategies for making the other set of data appear negative.

5. What type of graph will you use to represent this set of data? Why?

6. Draw the graph.

**Materials:** Calculator

**Explore:** Find a pattern for the sums.

$$2^0$$
$$2^0 + 2^1$$
$$2^0 + 2^1 + 2^2$$
$$2^0 + 2^1 + 2^2 + 2^3$$
$$\vdots$$

**1.** Evaluate the sums. Discuss any patterns you find.

**2.** Discuss possible strategies for finding a formula for finding these sums. Which strategy did your group choose? Why?

**3.** Find a formula for the sum $2^0 + 2^1 + \cdots + 2^n$.

**4.** Estimate $2^0 + 2^1 + 2^2 + 2^3 + 2^4 + 2^5 + 2^6$. Use your formula to find the sum. How did your estimate compare to the answer?

**Materials:** Calculator

**Explore:** Which situation would be more profitable?

    A: On September 1, you receive a penny. On each day following, you receive double what you were given the previous day. This continues for the 30 days of September.

    B: You are given $1,000,000.

**1.** Make a hypothesis.

**2.** Consider possible strategies for solving this problem. Which strategy did you choose? Why?

**3.** Which situation is more profitable?

**4.** What did you learn from the group activity that helped you answer this question?

**5.** Was it necessary to find the total amount received in situation A? Explain your answer.

**Materials:** None

**Explore:** Examine the products of numbers whose values are not known, but some properties of their factors are known.

- If $0 < a < 1$ and $0 < b < 1$, then approximate where on the number line $ab$ appears in relation to $a$ and $b$?

- If $a > 1$ and $b > 1$, then approximate where on the number line $ab$ appears in relation to $a$ and $b$?

- If $0 < a < 1$ and $b > 1$, then approximate where on the number line $ab$ appears in relation to $a$ and $b$?

1. Rewrite the problem in your own words.

2. Discuss possible plans for solving this problem. Which plan did your group choose? Why?

3. Use your plan to answer each question.

4. Did you change your plan for each question? If yes, which was the most efficient? If no, did the plan work better on a particular question?

**Materials:**   None

**Explore:**   Given the figure, determine which points on the number line
represent $E \cdot G$, $B \cdot F$, and $G \cdot I$.

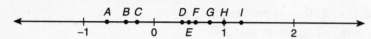

1. Outline a plan for solving the problem.

2. What information from the group activity can you use to solve this
problem?

3. Do you need any additional information? If so, what?

4. Use your plan to solve the problem.

## Chapter 8
## Group Assessment

*(Use after Lesson 8.7)*

**Names** _____

_____

**Materials:** None

**Explore:** Four people attend the auditions for three major male parts in the school play. How many different outcomes are possible?

**1.** Draw a tree diagram to determine the total number of possible outcomes.

**2.** Use the Counting Principle to determine the total number of possible outcomes.

**3.** Which method did you like the best? Explain your answer.

**Materials:** None

**Explore:** Sixteen people attend the auditions for the three major female parts in the school play. How many different outcomes are possible?

1. What method will you use to solve this problem?

2. How many possible outcomes are there?

3. What did you learn in the group activity that helped you solve this problem?

**Materials:**   Graph paper

**Explore:**   In the figure, what is the shortest path from *A* to *B* to *C*? You may move horizontally, vertically, or on a 45° diagonal.

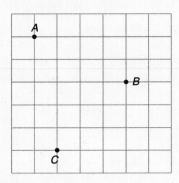

1. Discuss different plans for solving the problem. Which plan did your group choose? Why?

2. Use your plan to draw a shortest path.

3. Is there more than one shortest path?  If so, what do they have in common?

4. If the grid is made of 1 in. × 1 in. squares, what is the length of a shortest path?

**Materials:**   Graph paper

**Explore:**   In the figure, what is the shortest path from *A* to *B* to *C*? You may move horizontally, vertically, or on a 45° diagonal within the grids.

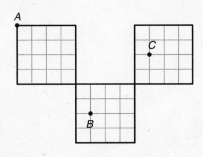

1. What plan will you use to solve this problem?

2. Use your plan to find a shortest path.

3. Is the answer unique? If not, what do all the shortest paths have in common?

4. If the grid is made of 1 in. × 1 in. squares, what is the length of the path?

5. If the grid were a solid 8 in. × 12 in. rectangle, would there be a shorter path? Explain your answer.

*(Use after Lesson 10.7)*

**Names** _____

_____

**Materials:** Ruler and paper

**Explore:** How many lines of symmetry does a regular *n*-gon have if *n* is odd?

**1.** Discuss strategies for solving this problem. Which strategy did your group select? Why?

**2.** Write a formula for the number of lines of symmetry, *l*, in a regular *n*-gon.

**3.** What do all lines of symmetry have in common?

**4.** How many lines of symmetry does a regular nonagon have? Sketch a regular nonagon and its lines of symmetry.

**Materials:**   Ruler and paper

**Explore:**   How many lines of symmetry are there in a regular *n*-gon?

1. What did you learn from the group activity that will help you solve this problem?

2. What method will you use to solve the problem? Is it the same as the method used in the group activity? Explain your decision.

3. What is the solution to the problem?

4. How many lines of symmetry are there in a regular decagon? Sketch a regular decagon and its lines of symmetry.

**Materials:** None

**Explore:** Given $\triangle ABC$, and $\overline{AB} \cong \overline{DE}$, $\overline{AC} \cong \overline{DF}$, and $\angle A \cong \angle D$, can you determine that $\triangle DEF$ is congruent to $\triangle ABC$?

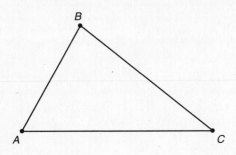

1. State a hypothesis.

2. Outline a plan to verify your hypothesis.

3. Is $\triangle DEF \cong \triangle ABC$? Was your hypothesis correct?

4. Does your answer depend on the given $\triangle ABC$, or given a different triangle and the corresponding information, is the same conclusion reached?

**Materials:** None

**Explore:** Given $\triangle ABC$, and $\overline{AB} \cong \overline{DE}$, $\overline{BC} \cong \overline{EF}$, and $\angle A \cong \angle D$, can you conclude that $\triangle ABC \cong \triangle DEF$?

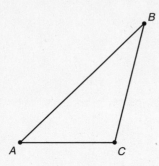

1. How is this problem different from the group activity?

2. State a hypothesis.

3. Outline a plan to verify your hypothesis.

4. Is $\triangle ABC \cong \triangle DEF$? Was your hypothesis correct? Explain your answer.

5. Does your answer depend on the given $\triangle ABC$?

**Names** _____

_____

**Materials:**   Small marshmallows and toothpicks

**Explore:**   Find a relationship between the number of vertices, faces, and edges of pyramids.

**1.** Construct three examples of pyramids with the marshmallows and toothpicks.

**2.** Complete the following table:

| Pyramid | Faces (F) | Vertices (V) | Edges (E) |
|---------|-----------|--------------|-----------|
| Example 1 | | | |
| Example 2 | | | |
| Example 3 | | | |

**3.** Outline a plan for finding the relationship between these numbers.

**4.** What is the relationship between the number of faces, vertices, and edges of a pyramid?

**Materials:**  Small marshmallows and toothpicks

**Explore:**  Find a relationship between the number of faces, vertices, and
edges of any polyhedron.

1. Make a hypothesis.

2. Determine a strategy for finding this relationship.

3. Did you use the same method as in the group activity? Why or
why not?

4. What is this relationship?

**Materials:**   Graph paper

**Explore:**   Find all pairs of numbers that satisfy all of the following
conditions: The sum of the two numbers is less than 10. The
sum of twice the first number and three times the second is
greater than 6. Both numbers are positive.

1. What strategy will your group use to solve this problem?

2. Is it possible to list all the pairs of numbers that satisfy the conditions?
   If so, list the pairs. If not, how will your group represent the pairs?

3. What pairs of numbers satisfy all of the conditions?

4. Explain why the pair 6 and 4 is not a solution.

5. Explain why the pair 1 and 1 is not a solution.

**Materials:**   Graph paper

**Explore:**   A publisher prints and binds paperback and hardcover books. It takes 10 minutes to print a paperback book and 30 minutes to print a hardcover book. It takes 30 minutes to bind a paperback book and 40 minutes to bind a hardcover book. The maximum amount of time per day spent on printing and binding is 900 minutes and 1600 minutes, respectively. Find all combinations of paperback and hardcover books that can be printed and bound in one day.

1.  Write an inequality that represents the number of books that can be printed in one day.

2.  Write an inequality that represents the number of books that can be bound in one day.

3.  Write an inequality that represents the fact that the number of paperbacks printed and bounded each day is not negative.

4.  Write an inequality that represents the fact that the number of hardcover books printed and bound each day is not negative.

5.  What information from the group activity will you use to solve this problem?

6.  Use a graph to show how many paperback and hardcover books can be printed and bound in one day.

**Materials:**   Calculator

**Explore:**   Here are two equations that contain quotients of polynomials.

$$y = \frac{2x^2 + 3x + 1}{x^2 - 4x + 3} \qquad y = \frac{x^3 - 2x + 1}{4x^3 + 2}$$

Examine what happens to the values of $y$ as the values of $x$ in the polynomials get extremely large.

**1.** Complete the table for the equation $y = \dfrac{2x^2 + 3x + 1}{x^2 - 4x + 3}$.

| $x$ | 10 | 100 | 1,000 | 10,000 | 1,000,000 |
|---|---|---|---|---|---|
| $y$ | | | | | |

**2.** Complete the table for $y = \dfrac{x^3 - 2x + 1}{4x^3 + 2}$.

| $x$ | 10 | 100 | 1,000 | 10,000 | 1,000,000 |
|---|---|---|---|---|---|
| $y$ | | | | | |

**3.** In each case, what does $y$ approach as $x$ gets larger and larger? (Write your answer as a fraction.)

**4.** Find a correlation between the coefficients of the polynomials that make up the equations and the values from Exercise 3.

**Materials:**  Calculator

**Explore:**  Examine the equations $y = \dfrac{3x^2 + 2x - 1}{x^2 + 1}$ and $y = \dfrac{2x^3 - x + 2}{4x^3 - 3}$.

Determine what happens to the value of $y$ when the values of $x$ get extremely small.

1. Complete the table for the equation $y = \dfrac{3x^2 + 2x - 1}{x^2 + 1}$.

| $x$ | $-10$ | $-100$ | $-1,000$ | $-10,000$ | $-1,000,000$ |
|-----|-------|--------|----------|-----------|--------------|
| $y$ |       |        |          |           |              |

2. Complete the table for the equation $y = \dfrac{2x^3 - x + 2}{4x^3 - 3}$.

| $x$ | $-10$ | $-100$ | $-1,000$ | $-10,000$ | $-1,000,000$ |
|-----|-------|--------|----------|-----------|--------------|
| $y$ |       |        |          |           |              |

3. In each case, what does $y$ approach as $x$ gets smaller and smaller? (Write your answer as a fraction.)

4. Find a correlation between the coefficients of the polynomials and the values from Exercise 3.

5. Without completing a table, what does the $y$-value of $y = \dfrac{7x^5 - 3x^2 + 6x + 2}{4x^5 + 3x^2 - 8}$ approach as $x$ gets extremely small?

# Answers to Chapter Projects

## ■ Chapter 1

*Cooperative Learning (Page 15)*

Answers vary.

Possible identifiers: route numbers, grid numbers, telephone numbers

Possible measurements: distance (miles and kilometers), map price, elevations

Possible sequences: grid numbers

*Cooperative Learning (Page 15)*

NOTE CARD 1: 4.12, 2.24, 3.16, 5.83
NOTE CARD 2: 1.95, 1.45, 2.70, 3.35
NOTE CARD 3: 0.71, 0.55, 0.95, 0.14
NOTE CARD 4: 0.71, 0.75, 0.87, 0.82
NOTE CARD 5: 1.83, 1.07, 2.29, 1.00

The square root of a number larger than 1 is smaller than the original number. The square root of a number between 0 and 1 is larger than the original number.

*Open-Ended Question (Page 16)*

a. $16 \div (4 + 4) + 3 = 5$

b. $12 \div (3 \cdot 2) + 7 = 9$ or $12 \div 3(2) + 7 = 9$

c. $(2 + 4) \div 2 \cdot 5 = 15$

d. $(2 + 3)^2 - 6 = 19$

*Oral Presentation (Page 16)*

Answers vary.

## ■ Chapter 2

*Cooperative Learning (Page 17)*

Answers vary.

*Experimentation (Page 17)*

Answers vary.

*Cooperative Learning (Page 18)*

3. a. $A = \frac{3}{4}x$

   b. broccoli: $\frac{9}{8}$ lb    bouillon: $\frac{3}{4}$ T
   onion: $\frac{3}{8}$ c    salt: $\frac{9}{16}$ t
   margarine: $\frac{3}{2}$ T    pepper: $\frac{3}{32}$ t
   flour: $\frac{3}{2}$ T    cream: $\frac{3}{8}$ c
   water: $\frac{27}{8}$ c

   c. $\frac{3}{8}$ c, $\frac{3}{2}$ T, $\frac{27}{8}$ c, and $\frac{3}{4}$ T

## ■ Chapter 3

*Cooperative Learning (Page 19)*

Answers vary.

*Problem Solving (Page 19)*

**1.** $-$    **2.** $-$    **3.** $+$    **4.** $-$

*Journal Entry (Page 20)*

a and b are correct models.

c is not correct. The equation states that the profit from the car wash minus the previously earned money equals the printer cost. The profit and previously earned money should be added.

*Cooperative Learning (Page 20)*

Answers vary.

## ■ Chapter 4

*Problem Solving (Page 21)*

a. $x = 2$

b. $x = -\frac{19}{7}$

c. $x = 7$

d. $x = \frac{2}{5}$

*Cooperative Learning (Page 21 and 22)*

6. a. 1840

   b. 1910

   c. 1880

   d. 1910

   e. 1840, 1850, 1860, and 1870

*Cooperative Learning (Page 22)*

3. 1 decimal place: 0.9; 0.8; 0.6; 0.4; 0.2; 0
   2 decimal places: 0.9; 0.81; 0.66; 0.44; 0.19; 0.04; 0
   3 decimal places: 0.9; 0.81; 0.656; 0.430; 0.185; 0.034; 0.001; 0
   4 decimal places: 0.9; 0.81; 0.6561; 0.4305; 0.1853; 0.0343; 0.0012; 0

**4. a.** 6; 7; 8; 8

   **b.** Answers vary. Possible answers: For the first iterations, the answers stayed the same. As more iterations were completed difference began to appear.

   **c.** Yes. Answers vary. Possible answers: Since previous answers are used to calculate the new numbers, any round off error will affect the new number.

**5.** 1 decimal place: 1.4; 1.2; 1.1; 1
2 decimal places: 1.41; 1.19; 1.09; 1.04; 1.02; 1.01; 1
3 decimal places: 1.414; 1.189; 1.090; 1.044; 1.022; 1.011; 1.005; 1.002; 1.001; 1
Answers may vary due to calculator rounding.
4 decimal places: 1.4142; 1.1892; 1.0905; 1.0443; 1.0219; 1.0109; 1.0054; 1.0027; 1.0013; 1.0006; 1.0003; 1.0001; 1

### ■ Chapter 5

*Research Project (Page 23)*

Answers vary.

*Journal Entry (Page 23)*

Answers vary.

*Interview (Page 23)*

Answers vary.

*Open-Ended Question (Page 24)*

Answers vary.

*Experiment (Page 24)*

Answers vary.

### ■ Chapter 6

*Math Game (Page 25)*

Numbers skipped from 1–100: 7, 14, 21, 27, 28, 35, 37, 42, 47, 49, 56, 57, 63, 67, 70, 71, 72, 73, 74, 75, 76, 77, 78, 79, 84, 87, 91, 97, 98

*Math Game (Page 25)*

Game varies.

*Cooperative Learning (Page 26)*

**3. a.** 1

   **b.** 2

   **c.** 32

   **d.** $n$th number $= 2^{n-1}$

   **e.** 524,288

**4. a.** 3; 2; 96; $3(2)^{n-1}$; 1,572,864

   **b.** 4; $\frac{1}{2}$; $\frac{1}{8}$; $4\left(\frac{1}{2}\right)^{n-1}$; $\frac{1}{131072}$

**5. a.** $n$th number

       $= $ (first number)(common ratio)$^{n-1}$

   **b.** $n$th number $= 5\left(\frac{2}{3}\right)^{n-1}$

### ■ Chapter 7

*Journal Entry (Page 27)*

Answers vary.

*Journal Entry (Page 27)*

No. Answers vary.

*Cooperative Learning (Page 27)*

   **a.** Answers vary.

   **b.**

| Money Bet | Money Returned | Profit/Loss |
|---|---|---|
| $10 | $ 9.80 | $ 0.20 loss |
| $20 | $19.60 | $ 0.40 loss |
| $30 | $29.40 | $ 0.60 loss |
| $40 | $39.20 | $ 0.80 loss |
| $50 | $49.00 | $ 1.00 loss |
| $100 | $98.00 | $ 2.00 loss |
| $1000 | $980.00 | $20.00 loss |

   **c.** Lose

   **d.** Over 100% return

   **e.** Answers vary.

*Cooperative Learning (Page 28)*

**3. a.**

| Year | Initial Bal. | Interest Earned | Ending Bal. |
|---|---|---|---|
| 1 | $200.00 | $ 8.50 | $208.50 |
| 2 | $308.50 | $13.11 | $321.61 |
| 3 | $421.61 | $17.92 | $439.53 |
| 4 | $539.53 | $22.93 | $562.46 |

   **b.** Scenario III is the most profitable.

   **d.** Always; always; always; sometimes

■ **Chapter 8**

*Journal Entry (Page 29)*

Answers vary.

*Demonstration (Page 29)*

Answers vary.

*Cooperative Learning (Page 29)*

**3. a.** Answers vary.

**b.** 360 ft × 160 ft; Area = 57600 ft$^2$

**c.** Answers vary. ≈ 3 ft$^2$

**d.**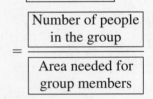

**e.** Answers vary.

**f.** Answers vary.

*Research Project (Page 30)*

Answers vary.

*Experiment (Page 30)*

**1.** $\frac{1}{2}$

**2.–6.** Answers vary.

**7.** The more data that is used, the closer the experimental probability gets to the theoretical probability.

■ **Chapter 9**

*Cooperative Learning (Page 31)*

Answers vary.

*Construction (Page 31)*

**3.** $\sqrt{2}$    **4.** $\sqrt{3}$

**5.** 2; $\sqrt{5}$; $\sqrt{6}$; $\sqrt{7}$; 2$\sqrt{2}$; 3; $\sqrt{10}$; $\sqrt{11}$; 2$\sqrt{3}$; $\sqrt{13}$; $\sqrt{14}$; $\sqrt{15}$

**6.** The measurements are the same as the decimal approximations of exercise 5.

**7.** Nautilus Shell

*Cooperative Learning (Page 32)*

Answers vary.

*Cooperative Learning (Page 32)*

**6.** $2s + 2 < 3s + 1$    **7.** $1 < s$

**8.** There is more than one penny in the stack.

**9. a.** $4s + 3 > 2s + 5 \rightarrow s > 1$; There is more than one penny in the stack.

**b.** $s + 6 < 4s \rightarrow 2 < s$; There is more than two pennies in the stack.

**c.** $3s + 2 = 2s + 5 \rightarrow s = 3$; There are 3 pennies in the stack.

■ **Chapter 10**

*Cooperative Learning (Page 33)*

**3.**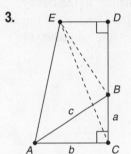

**4. a.** $\triangle AEC$ and $\triangle EDC$

**b.** $a + b$

**c.** They are the same.

**d.** $\frac{1}{2}a^2 + \frac{1}{2}ab$

**e.** $\frac{1}{2}b^2 + \frac{1}{2}ab$

**f.** $\frac{1}{2}a^2 + \frac{1}{2}b^2 + ab$

**5. a.** $\triangle ABC$, $\triangle ABE$, and $\triangle EDB$

**b.** c

**c.** $\frac{1}{2}ab$

**d.** $\frac{1}{2}c^2$

**e.** $\frac{1}{2}ab$

**f.** $ab + \frac{1}{2}c^2$

**7.** $a^2 + b^2 = c^2$ Pythagorean Theorem

*Research Project (Page 34)*

Symmetric: a, b, d, g, h, and i

Not symmetric: c, e, f, and j

*Construction (Page 34)*

**3. a.** Isosceles triangles are symmetric with respect to a line that passes through the vertex where the equal sides meet and the middle of the base.

**b.** The base of the right triangle is half the length of the base of the isosceles triangle.

**4. a.** Infinitely many.

**b.** The third vertices lie directly on the line of symmetry.

**c.** To draw an isosceles triangle you need to draw a line perpendicular to the base through its midpoint. Then, place the third vertex anywhere on this line and draw the two congruent sides.

**5.**

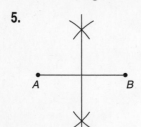

**6.** Answers vary.

■ **Chapter 11**

*Cooperative Learning (Page 35)*

**3.** 136.5 units$^2$; Answers vary.

**4.** Answers vary.

*Cooperative Learning (Page 35)*

**3. a.** $\frac{1}{2}$; $\frac{2}{3}$; $\frac{1}{2}$; $\frac{2}{3}$

**b.** 12 and 48; 10 and $\frac{45}{2}$; 3 and 12; 14 and $\frac{63}{2}$

**c.** $\frac{1}{4}$; $\frac{4}{9}$; $\frac{1}{4}$; $\frac{4}{9}$

**d.** (ratio of areas) = (scale factor)$^2$

**5.** All answers lead to (ratio of areas) = (scale factor)$^2$

■ **Chapter 12**

*Problem Solving (Page 37)*

**1.** The sign alternates from + to −.

**2.** Numerator is always 1.

**3.** Denominators are odd numbers.

**4.** c; This expression follows the three patterns from the previous questions.

**5.** When $n = 6$, $\pi \approx 2.976046176$. When $n = 15$, $\pi \approx 3.20818$. When $n = 30$, $\pi \approx 3.1082629$.

**6.** Error: 0.1655 ($n = 6$); 0.0666 ($n = 15$); 0.0333 ($n = 30$)

*Journal Entry (Page 37)*

**1.** A capsule is a cylinder with a hemisphere on each end. A corner curio cabinet is a quarter of a cylinder.

**2.** Capsule: $V = \frac{4}{3}\pi r^3 + Bh$; Curio: $B = \frac{1}{4}Bh$

**3.** Answers vary.

**4.** Answers vary.

*Problem Solving (Page 38)*

**3.** SPHERE: **a.** $\approx 13.4$ ft
**b.** $\approx 2256.4$ ft$^2$

CYLINDER: **a.** $10,000 = \pi r^2 h$
**b.**

| $h$ | 39.3 | 31.8 | 26.3 | 22.1 | 18.8 | 16.2 |
|---|---|---|---|---|---|---|
| $S$ | 2731.3 | 2626.4 | 2578.0 | 2571.1 | 2597.5 | 2656.5 |

**c.** $r = 12$ ft and $h = 22.1$ ft; $S = 2571.1$ ft$^2$

PRISM: **a.** $10,000 = Bh = x^2 h$
**b.**

| $h$ | 27.7 | 25 | 22.7 | 20.7 | 18.9 | 17.4 |
|---|---|---|---|---|---|---|
| $S$ | 2827.2 | 2800 | 2788.8 | 2789.6 | 2796.8 | 2822.4 |

**c.** $r = 21$ ft and $h = 22.7$ ft; $S = 2788.8$ ft$^2$

**4.** Sphere

■ **Chapter 13**

*Graphing Calculator (Page 38)*

**1.** j.; $y = \frac{1}{2}x + 3$

**4.** $y \approx 0.315x - 1.347$

*Graphing Calculator (Page 38)*

**3. a.** $(-2, -4)$

**c.** No; Lines that cross can do so only at one point.

**4. a.** Yes   **b.** Yes

**c.** Both lines have slope 2.

**5. a.** Yes

   **b.** After zooming out several times, the lines appear to intersect.

   **c.** Their slopes are not the same.

*Cooperative Learning (Page 39)*

Answers vary.

■ **Chapter 14**

*Journal Entry (Page 40)*

Sets 3 and 4; Answers vary but should include the idea that the sets with a few extreme terms usually use the median as the measure of central tendency.

*Cooperative Learning (Page 40)*

**3. a.** $(3, 7)$   **b.** $(-1, -1)$   **c.** $\left(\frac{11}{7}, \frac{5}{14}\right)$

**4.** Division by 0 occurs

*Graphing Calculator (Page 41)*

**3. a.** Answers vary.

   **b.** They are horizontal lines.

**4. a.** Answers vary.

   **b.** They are lines which are not horizontal or vertical.

**5.** They are parabolas (U-shaped).

*Cooperative Learning (Page 42)*

**3. b.** $x^2 + 2x + 1$

   **c.** $x^3 + 3x^2 + 3x + 1$

   **d.** $x^4 + 4x^3 + 6x^2 + 4x + 1$

**4.** 1, 3, 3, and 1; 1, 4, 6, 4, and 1

**5.** Pascal's Triangle

**6.** $x^8 + 8x^7 + 28x^6 + 56x^5 + 70x^4 + 56x^3$
   $+ 28x^2 + 8x + 1$

# Answers to Partner Quizzes

## ■ Chapter 1 (Page 44)

1. 47, 95, 191

2.

3. ÷, /, fraction bar

4. $12 - 7 = 5$   5. 214.25

6. 64 in.³   7. 60.8 ft   8. 6

9. $(21 \div 3 + 4) \times 2$

10. $53.50

## ■ Chapter 2 (Page 45)

1. $2x + 2y + 10$   2. 24 points

3. $13a + 7$   4. $9x + 3$; 21 in.²

5. b. $x = 3$   6. 4   7. $6   8. 7.153

9. 12 years   10. $3.83

## ■ Chapter 3 (Page 46)

1.

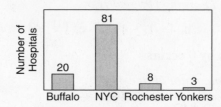

2. >   3. 0   4. −7

5. $5x - 8$; 12   6. $3 - 4 + 2 - 1 = x$; 0

7. $b + (-a)$   8. Sometimes   9. −9

10. Never

## ■ Chapter 4 (Page 47)

1. Multiplication

2. −14   3. $2x + 9 = 13$; 2   4. −2

5. 28°   6. $-\frac{3}{2}$   7. 1:35 P.M.

8. $4(x + 2) = 12$; 1   9. −1

10. $2(x + 1) + x = 38$ or $3x + 2 = 38$; 12 in.

## ■ Chapter 5 (Page 48)

1. Answers vary. ≈ 596,270

2. Answers vary. ≈ 1,003,464

3. Providence

4.

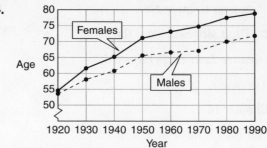

5. Increasing   6. 1920

7. Bar graph or picture graph

## ■ Chapter 6 (Page 49)

1. 1, 2, 3, 4, 6, 11, 12, 22, 33, 44, 66, 132

2. 4 in. × 3 in. × 5 in., 6 in. × 2 in. × 5 in.,
   10 in. × 2 in. × 3 in., 15 in. × 2 in. × 2 in.,
   1 in. × 2 in. × 30 in., 1 in. × 3 in. × 20 in.,
   1 in. × 4 in. × 15 in., 1 in. × 5 in. × 12 in.,
   1 in. × 6 in. × 10 in.

3. $2^2 \cdot 3^2 \cdot 5 \cdot 7$

4. Sometimes   5. $4x^2y^2$

6. Answers vary. Examples: 49, 77, 91

7. $24 \, x^2y^3$

8. Answers vary. Examples: $\frac{4}{5}, \frac{24}{30}, \frac{36}{45}$

9. $\frac{56}{46}$   10. $\frac{5}{11}, \frac{4}{9}, \frac{9}{22}, \frac{2}{5}$

## ■ Chapter 7 (Page 50)

1. $\frac{3}{8} + \frac{7}{8} - \frac{2}{8}$; 1   2. $\frac{2a}{b}$   3. $\frac{9}{8}$ or $1\frac{1}{8}$

4. $-\frac{7}{15}$   5. Sometimes   6. 0.373

7. $\frac{1}{4}$   8. $-\frac{3}{10}$   9. It increases.

10. $11\frac{5}{8}$ or 11.625

**■ Chapter 8 (Page 51)**

1. Ratio    2. 6.4 oz tube    3. $\frac{10}{10}$

4. $a = \frac{10}{3}$    5. 3    6. 1.68 ft

7. 415 words    8. 3.12

9. $\approx 14\%$    10. 75 games

**■ Chapter 9 (Page 52)**

1. $x^2 - 4 = 8$; 3.464, $-3.464$    2. 14 in.

3. $\frac{4}{5}$    4. Rational    5. 24    6. 39

7. $i^2 = h^2 + g^2$    8. 4.6 miles

9. 11.6 miles    10. 4.6 square miles

**■ Chapter 10 (Page 53)**

1. B    2. $\overleftrightarrow{AB}$ and $\overleftrightarrow{CD}$

3. $\angle BEC$ and $\angle CED$

4. $\angle CEA$    5. 148°

6. $\angle 3, \angle 6, \angle 8, \angle 9, \angle 11, \angle 14,$ and $\angle 16$

7.

8. Answers vary.    9. 19° each

10. $m\angle A = m\angle B = m\angle C = 60°$

**■ Chapter 11 (Page 54)**

1. $5\sqrt{7} \approx 13.23$    2. $10 + 4\sqrt{2} \approx 15.66$

3.

4. Answers vary.

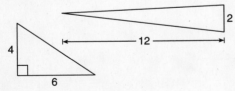

5.
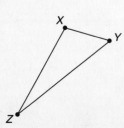

6. $A'(0, 2), B'(-3, 1), C'(-4, 4)$

7. Same: I, II, III, V, X
   Turns into another numeral:
   IV → VI, VI → IV, IX → XI

8. $A'(-1, 0), B'(-4, 0), C'(-2, -2)$

9. $A'(-1, 0), B'(1, 1), C'(3, 0), D'(0, -3)$

10. $(x - 3, y + 4)$

**■ Chapter 12 (Page 55)**

1. 6 in.    2. $12\pi$ in. $\approx 37.68$ in.

3. $\approx 2.23$ in.

4. $(6\pi + 12)$ in. $\approx 30.84$ in.

5. 5; pentagon    6. Cylinder and cone

7. a.    8. 36 in.$^2$

9. $V = s^3$    10. 10 in.

**■ Chapter 13 (Page 56)**

1. Answers vary. $(2, 1), \left(0, \frac{5}{2}\right), (-2, 4)$

2. $x + 10 = y$

3. The $x$-coordinates are all 4.

4.
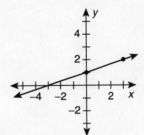

5. $\left(\frac{15}{2}, 0\right)$    6. $(0, 0)$

7. The time when the solution reaches 0° C.

8. $m = \frac{1}{8}$    9. No    10. $m = -\frac{2}{3}$

**■ Chapter 14 (Page 57)**

1. 9|7 means 97.

| 9 | 7 5 2 |
|---|-------|
| 8 | 9 5 5 1 0 |
| 7 | 7 2 1 |
| 6 | 9 4 |
| 5 | 1 |
| 0 | 5 |

**2.** 74.2    **3.** 80

**4.** 85    **5.** Median

**6.** Answers vary. Possible answer:
65, 68, 71, 71, 74, 76, 79, 86, 93, 100
65, 68, 71, 71, 74, 76, 79, 86, 86, 100

**7.**

|  | High Temp | Low Temp |
|---|---|---|
| Monday | 93 | 72 |
| Tuesday | 90 | 70 |
| Wednesday | 88 | 75 |
| Thursday | 94 | 74 |
| Friday | 89 | 69 |
| Saturday | 85 | 66 |
| Sunday | 87 | 65 |

**8.** $\begin{bmatrix} 7 & 1 & -2 \\ 1 & -3 & 4 \\ -2 & -4 & 5 \end{bmatrix}$

# Answers to Group Assessment

■ **Chapter 1** (Page 59)

1. Answers vary.  **2.** 16, 64  **3.** $4^n$

4. Drawing the figures becomes difficult after only a few iterations. Other answers may vary.

■ **Chapter 1** (Page 60)

1. Answers vary.  **2.** 25, 125  **3.** $5^n$

4. The same principle is used. The difference is that for every segment, the new figure has 5 segments instead of 4 as in the group activity.

■ **Chapter 2** (Page 61)

1. Answers vary.  **2.** Answers vary.

3. $240 \leq 220 + x$ and $269.7 > 220 + x$

4. 20–49

■ **Chapter 2** (Page 62)

1. Answers vary.  **2.** Answers vary.

3. $320 \leq 290 + x$ and $359.6 > 290 + x$

4. 30–69

5. Yes. You must earn at least 70 points on the last test.

6. Answers vary.

■ **Chapter 3** (Page 63)

1. Quadrants 1 and 2  **2.** Answers vary.

3. V-shape with the point at the origin

4. The $y$-coordinates are the same. Opposites have the same absolute value.

■ **Chapter 3** (Page 64)

1. Quadrants 1 and 2  **2.** Answers vary.

3. V-shape with the point at $(1, 0)$

4. No. The $x$ values that are equal distances from 1 have the same $y$ values.

■ **Chapter 4** (Page 65)

1.
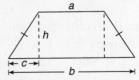

2. $A = \frac{1}{2}ch$; Yes, the triangles are the same size.

---

3. $A = ah$

4. $\boxed{\text{Area of isosceles trapezoid}} = 2 \boxed{\text{Area of triangle}}$ $+ \boxed{\text{Area of rectangle}}$

$A = ch + ah$ or $A = h(c + a)$

5. $b - a$ gives the length of both bases of the triangles. Thus, $\frac{1}{2}(b - a)$ is the base of one triangle (previously labeled $c$).

6. $A = \frac{1}{2}h(a + b)$

7. The length of the rectangle is $\frac{1}{2}(a + b)$ and the height is $h$. $A = \frac{1}{2}h(a + b)$; Answers vary.

■ **Chapter 4** (Page 66)

1.
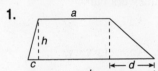

2. $A = \frac{1}{2}ch$, $A = \frac{1}{2}dh$; No, the triangles are different.

3. $A = ah$

4. $\boxed{\text{Area of trapezoid}} = \boxed{\text{Area of Triangle 1}}$ $+ \boxed{\text{Area of Triangle 2}} + \boxed{\text{Area of rectangle}}$

$A = \frac{1}{2}ch + \frac{1}{2}dh + ah$ or $A = h(\frac{1}{2}c + \frac{1}{2}d + a)$

5. Side $b$ has the same length as side $a$ plus the bases of the two triangles.

6. $c = b - a - d$; $A = \frac{1}{2}h(a + b)$

7. The formulas are the same.

8. No, the cut and rearrange method only worked because the triangles were the same size. You would not get a rectangle.

■ **Chapter 5** (Page 67)

1. Unemployment rate; The unemployment rate has been dropping while your candidate has been in office. In fact it is below the state's rate.

2. Answers vary.  **3.** Answers vary.

**4.** Answers vary. Use a broken vertical scale.

**5.** Answers vary.    **6.** Answers vary.

## ■ Chapter 5 (Page 68)

**1.** Although the number of crimes per 100,000 people has dropped during the incumbent's term, it has dropped very slowly and is still well above the state's number.

**2.** Answers vary.    **3.** Answers vary.

**4.** Answers vary. Use large intervals on the vertical axis.

**5.** Answers vary.    **6.** Answers vary.

## ■ Chapter 6 (Page 69)

**1.** 1, 3, 7, 15; Answers vary.

**2.** Answers vary.    **3.** Sum $= 2^{n+1} - 1$

**4.** Answers vary; 127; answers vary.

## ■ Chapter 6 (Page 70)

**1.** Answers vary.    **2.** Answers vary.

**3.** A    **4.** Answers vary.

**5.** No. Possible answers: On September 30 alone, you receive $5,368,709.12.

## ■ Chapter 7 (Page 71)

**1.** Answers vary.    **2.** Answers vary.

**3.** $ab > 0$ but $ab$ is less than both $a$ and $b$.

$ab > 1$ and greater than $a$ and $b$.

$ab > 0$, greater than $a$, but less than $b$.

**4.** Answers vary.

## ■ Chapter 7 (Page 72)

**1.** Answers vary.

**2.** $0 < a < 1$ and $0 < b < 1$ implies $ab > 0$ but is less than $a$ and $b$.

$0 < a < 1$ and $b > 1$ implies $ab > 0$, greater than $a$, but less than $b$.

**3.** If $-1 < a < 0$ and $0 < b < 1$, then $ab < 0$ and lies between $a$ and $0$.

**4.** $E \cdot G = D; B \cdot F = C; G \cdot I = H$

## ■ Chapter 8 (Page 73)

**1.**

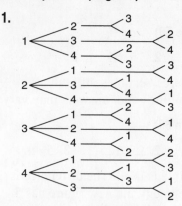

**2.** 24    **3.** Answers vary.

## ■ Chapter 8 (Page 74)

**1.** Answers vary.

**2.** 3360    **3.** Answers vary.

## ■ Chapter 9 (Page 75)

**1.** Answers vary.

**2.** Answers vary. Example:

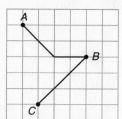

**3.** Yes; All shortest paths involve 2 diagonal and 2 horizontal steps to get to B and 3 diagonal steps to get to C.

**4.** $\left(2 + 5\sqrt{2}\right)$ in. $\approx 9.07$ in.

## ■ Chapter 9 (Page 76)

**1.** Answers vary.

**2.** Answers vary. Example:

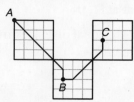

**3.** No. Moving from A to B involves 5 diagonal and 1 vertical move. Moving from B to C involves 1 horizontal, 1 vertical and 3 diagonal moves.

**4.** $\left(8\sqrt{2} + 3\right)$ in. $\approx 14.31$ in.

**5.** Yes. Moving from A to B would be the same. However, you could move from B to C in 4 diagonal steps. Moving one extra diagonal step is less than moving the vertical and horizontal step required to maneuver the corner.

■ **Chapter 10 (Page 77)**

**1.** Answers vary.    **2.** $l = n$

**3.** All lines of symmetry pass through a vertex and the midpoint of the opposite side.

**4.** 9;

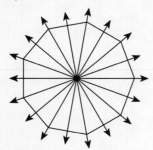

■ **Chapter 10 (Page 78)**

**1.** Answers vary.    **2.** Answers vary.

**3.** $l = n$

**4.** 10;

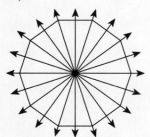

■ **Chapter 11 (Page 79)**

**1.** Answers vary.    **2.** Answers vary.

**3.** Yes; answers vary.

**4.** No. If you know the measure of two sides of a triangle and the angle between them, there is only one way to complete the triangle.

■ **Chapter 11 (Page 80)**

**1.** $\overline{BC} \cong \overline{EF}$ instead of $\overline{AC} \cong \overline{DF}$.

**2.** Answers vary.    **3.** Answers vary.

**4.** Not always; Answers vary;

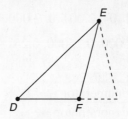 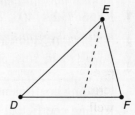

**5.** Yes. If $\angle A$ and $\angle D$ were obtuse, then you could guarantee that $\triangle ABC \approx \triangle DEF$.

■ **Chapter 12 (Page 81)**

**1.** Constructions vary.    **2.** Answers vary.

**3.** Answers vary.

**4.** Answers vary. Possible relationships: $V = F$, $E = 2F - 2 = 2V - 2$ or $V + F = E + 2$

■ **Chapter 12 (Page 82)**

**1.** Answers vary.    **2.** Answers vary.

**3.** Answers vary.    **4.** $V + F = E + 2$

■ **Chapter 13 (Page 83)**

**1.** Answers vary.

**2.** No. Shaded region.

**3.**

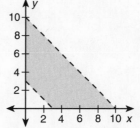

**4.** $6 + 4 = 10$, The sum must be less than 10.

**5.** $2(1) + 3(1) \not> 6$

■ **Chapter 13 (Page 84)**

**1.** $10x + 30y \leq 900$ or $y \leq -\frac{1}{3}x + 30$

**2.** $30x + 40y \leq 1600$ or $y \leq -\frac{3}{4}x + 40$

**3.** $x \geq 0$    **4.** $y \geq 0$    **5.** Answers vary.

**6.**

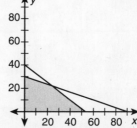

## ■ Chapter 14 (Page 85)

1. 3.66667, 2.11403, 2.01104, 2.00110, 2.00000

2. 0.24513, 0.24995, 0.25000, 0.25000, 0.25

3. $2; \frac{1}{4}$

4. $\dfrac{\text{leading coefficient of numerator}}{\text{leading coefficient of denominator}}$

## ■ Chapter 14 (Page 86)

1. 2.762376238, 2.97960204, 2.997996002, 2.99979996, 2.99998

2. 0.4966275294, 0.499974125, 0.4999997491, 0.4999999975, 0.5

3. $3, \frac{1}{2}$

4. $\dfrac{\text{leading coefficient of numerator}}{\text{leading coefficient of denominator}}$

5. $\frac{7}{4}$